Getting Started with Arduino

Massimo Banzi

First Edition

O'REILLY

BEIJING · CAMBRIDGE · FARNHAM · KÖLN · SEBASTOPOL · TAIPEI · TOKYO

Getting Started with Arduino
by Massimo Banzi

Published by Make:Books, an imprint of Maker Media,
a division of O'Reilly Media, Inc.
1005 Gravenstein Highway North, Sebastopol, CA 95472

O'Reilly books may be purchased for educational, business,
or sales promotional use. For more information, contact our
corporate/institutional sales department: 800-998-9938
or corporate@oreilly.com.

Print History: October 2008: First Edition

Publisher: Dale Dougherty
Associate Publisher: Dan Woods
Executive Editor: Brian Jepson
Creative Director: Daniel Carter
Designer: Brian Scott
Production Manager: Terry Bronson
Copy Editor: Nancy Kotary
Indexer: Patti Schiendelman
Illustrations: Elisa Canducci

The O'Reilly logo is a registered trademark of O'Reilly Media, Inc.
The Make: Projects series designations and related trade dress
are trademarks of O'Reilly Media, Inc. The trademarks of third
parties used in this work are the property of their respective
owners.

Important Message to Our Readers: Your safety is your own
responsibility, including proper use of equipment and safety gear,
and determining whether you have adequate skill and experience. Electricity and other resources used for these projects are
dangerous unless used properly and with adequate precautions,
including safety gear. Some illustrative photos do not depict
safety precautions or equipment, in order to show the project
steps more clearly. These projects are not intended for use by
children.

Use of the instructions and suggestions in *Getting Started with
Arduino* is at your own risk. O'Reilly Media, Inc. and the author
disclaim all responsibility for any resulting damage, injury, or
expense. It is your responsibility to make sure that your activities
comply with applicable laws, including copyright.

ISBN: 978-0-596-15551-3   [3/10]

Contents

Preface

A few years ago I was given a very interesting challenge: teach designers the bare minimum in electronics so that they could build interactive prototypes of the objects they were designing.

I started following a subconscious instinct to teach electronics the same way I was taught in school. Later on I realised that it simply wasn't working as well as I would like, and started to remember sitting in a class, bored like hell, listening to all that theory being thrown at me without any practical application for it.

In reality, when I was in school I already knew electronics in a very empirical way: very little theory, but a lot of hands-on experience.

I started thinking about the process by which I really learned electronics:

» I took apart any electronic device I could put my hands on.

» I slowly learned what all those components were.

» I began to tinker with them, changing some of the connections inside of them and seeing what happened to the device: usually something between an explosion and a puff of smoke.

» I started building some kits sold by electronics magazines.

» I combined devices I had hacked, and repurposed kits and other circuits that I found in magazines to make them do new things.

As a little kid, I was always fascinated by discovering how things work; therefore, I used to take them apart. This passion grew as I targeted any unused object in the house and then took it apart into small bits. Eventually, people brought all sorts of devices for me to dissect. My biggest projects at the time were a dishwasher and an early computer that came from an insurance office, which had a huge printer, electronics cards, magnetic card readers, and many other parts that proved very interesting and challenging to completely take apart.

After quite a lot of this dissecting, I knew what electronic components were and roughly what they did. On top of that, my house was full of old electronics magazines that my father must have bought at the beginning of the 1970s. I spent hours reading the articles and looking at the circuit diagrams without understanding very much.

This process of reading the articles over and over, with the benefit of knowledge acquired while taking apart circuits, created a slow virtuous circle.

A great breakthrough came one Christmas, when my dad gave me a kit that allowed teenagers to learn about electronics. Every component was housed in a plastic cube that would magnetically snap together with other cubes, establishing a connection; the electronic symbol was written on top. Little did I know that the toy was also a landmark of German design, because Dieter Rams designed it back in the 1960s.

With this new tool, I could quickly put together circuits and try them out to see what happened. The prototyping cycle was getting shorter and shorter.

After that, I built radios, amplifiers, circuits that would produce horrible noises and nice sounds, rain sensors, and tiny robots.

I've spent a long time looking for an English word that would sum up that way of working without a specific plan, starting with one idea and ending up with a completely unexpected result. Finally, "tinkering" came along. I recognised how this word has been used in many other fields to describe a way of operating and to portray people who set out on a path of exploration. For example, the generation of French directors who gave birth to the "Nouvelle Vague" were called the "tinkerers". The best definition of tinkering that I've ever found comes from an exhibition held at the Exploratorium in San Francisco:

Tinkering is what happens when you try something you don't quite know how to do, guided by whim, imagination, and curiosity. When you tinker, there are no instructions—but there are also no failures, no right or wrong ways of doing things. It's about figuring out how things work and reworking them.

Contraptions, machines, wildly mismatched objects working in harmony—this is the stuff of tinkering.

Tinkering is, at its most basic, a process that marries play and inquiry.

—www.exploratorium.edu/tinkering

From my early experiments I knew how much experience you would need in order to be able to create a circuit that would do what you wanted starting from the basic components.

Another breakthrough came in the summer of 1982, when I went to London with my parents and spent many hours visiting the Science Museum. They had just opened a new wing dedicated to computers, and by following a series of guided experiments, I learned the basics of binary math and programming.

There I realised that in many applications, engineers were no longer building circuits from basic components, but were instead implementing a lot of the intelligence in their products using microprocessors. Software was replacing many hours of electronic design, and would allow a shorter tinkering cycle.

When I came back I started to save money, because I wanted to buy a computer and learn how to program.

My first and most important project after that was using my brand-new ZX81 computer to control a welding machine. I know it doesn't sound like a very exciting project, but there was a need for it and it was a great challenge for me, because I had just learned how to program. At this point, it became clear that writing lines of code would take less time than modifying complex circuits.

Twenty-odd years later, I'd like to think that this experience allows me to teach people who don't even remember taking any math class and to infuse them with the same enthusiasm and ability to tinker that I had in my youth and have kept ever since.

—Massimo

Acknowledgments

This book is dedicated to Luisa and Alexandra.

First of all I want to thank my partners in the Arduino Team: David Cuartielles, David Mellis, Gianluca Martino, and Tom Igoe. It is an amazing experience working with you guys.

Barbara Ghella, she doesn't know but, without her precious advice, Arduino and this book might have never happened.

Bill Verplank for having taught me more than Physical Computing.

Gillian Crampton-Smith for giving me a chance and for all I have learned from her.

Hernando Barragan for the work he has done on Wiring.

Brian Jepson for being a great editor and enthusiastic supporter all along.

Nancy Kotary, Brian Scott, Terry Bronson, and Patti Schiendelman for turning what I wrote into a finished book.

I want to thank a lot more people but Brian tells me I'm running out of space so I'll just list a small number of people I have to thank for many reasons:

Adam Somlai-Fisher, Ailadi Cortelletti, Alberto Pezzotti, Alessandro Germinasi, Alessandro Masserdotti, Andrea Piccolo, Anna Capellini, Casey Reas, Chris Anderson, Claudio Moderini, Clementina Coppini, Concetta Capecchi, Csaba Waldhauser, Dario Buzzini, Dario Molinari, Dario Parravicini, Donata Piccolo, Edoardo Brambilla, Elisa Canducci, Fabio Violante, Fabio Zanola, Fabrizio Pignoloni, Flavio Mauri, Francesca Mocellin, Francesco Monico, Giorgio Olivero, Giovanna Gardi, Giovanni Battistini, Heather Martin, Jennifer Bove, Laura Dellamotta, Lorenzo Parravicini, Luca Rocco, Marco Baioni, Marco Eynard, Maria Teresa Longoni, Massimiliano Bolondi, Matteo Rivolta, Matthias Richter, Maurizio Pirola, Michael Thorpe, Natalia Jordan, Ombretta Banzi, Oreste Banzi, Oscar Zoggia, Pietro Dore, Prof Salvioni, Raffaella Ferrara, Renzo Giusti, Sandi Athanas, Sara Carpentieri, Sigrid Wiederhecker, Stefano Mirti, Ubi De Feo, Veronika Bucko.

How to Contact Us

We have verified the information in this book to the best of our ability, but you may find things that have changed (or even that we made mistakes!). As a reader of this book, you can help us to improve future editions by sending us your feedback. Please let us know about any errors, inaccuracies, misleading or confusing statements, and typos that you find anywhere in this book.

Please also let us know what we can do to make this book more useful to you. We take your comments seriously and will try to incorporate reasonable suggestions into future editions.

You can write to us at:

Maker Media
1005 Gravenstein Highway North
Sebastopol, CA 95472
(800) 998-9938 (in the U.S. or Canada)
(707) 829-0515 (international/local)
(707) 829-0104 (fax)

Maker Media is a division of O'Reilly Media devoted entirely to the growing community of resourceful people who believe that if you can imagine it, you can make it. Consisting of MAKE magazine, CRAFT magazine, Maker Faire, as well as the Hacks, Make:Projects, and DIY Science book series, Maker Media encourages the Do-It-Yourself mentality by providing creative inspiration and instruction.

For more information about Maker Media, visit us online:
MAKE   www.makezine.com
CRAFT: www.craftzine.com
Maker Faire: www.makerfaire.com
Hacks: www.hackszine.com

To comment on the book, send email to bookquestions@oreilly.com.

The O'Reilly web site for *Getting Started with Arduino* lists examples, errata, and plans for future editions. You can find this page at www.makezine.com/getstartedarduino.

For more information about this book and others, see the O'Reilly web site: www.oreilly.com.

For more information about Arduino, including discussion forums and further documentation, see www.arduino.cc.

1/Introduction

Arduino is an open source physical computing platform based on a simple input/output (I/O) board and a development environment that implements the Processing language (**www.processing.org**). Arduino can be used to develop standalone interactive objects or can be connected to software on your computer (such as Flash, Processing, VVVV, or Max/MSP). The boards can be assembled by hand or purchased preassembled; the open source IDE (Integrated Development Environment) can be downloaded for free from **www.arduino.cc**.

Arduino is different from other platforms on the market because of these features:

» It is a multiplatform environment; it can run on Windows, Macintosh, and Linux.

» It is based on the Processing programming IDE, an easy-to-use development environment used by artists and designers.

» You program it via a USB cable, not a serial port. This feature is useful, because many modern computers don't have serial ports.

» It is open source hardware and software—if you wish, you can download the circuit diagram, buy all the components, and make your own, without paying anything to the makers of Arduino.

- » The hardware is cheap. The USB board costs about €20 (currently, about US$35) and replacing a burnt-out chip on the board is easy and costs no more than €5 or US$4. So you can afford to make mistakes.

- » There is an active community of users, so there are plenty of people who can help you.

- » The Arduino Project was developed in an educational environment and is therefore great for newcomers to get things working quickly.

This book is designed to help beginners understand what benefits they can get from learning how to use the Arduino platform and adopting its philosophy.

Intended Audience

This book was written for the "original" Arduino users: designers and artists. Therefore, it tries to explain things in a way that might drive some engineers crazy. Actually, one of them called the introductory chapters of my first draft "fluff". That's precisely the point. Let's face it: most engineers aren't able to explain what they do to another engineer, let alone a regular human being. Let's now delve deep into the fluff.

NOTE: Arduino builds upon the thesis work Hernando Barragan did on the Wiring platform while studying under Casey Reas and me at IDII Ivrea.

After Arduino started to become popular, I realised how experimenters, hobbyists, and hackers of all sorts were starting to use it to create beautiful and crazy objects. I realised that you're all artists and designers in your own right, so this book is for you as well.

Arduino was born to teach Interaction Design, a design discipline that puts prototyping at the centre of its methodology. There are many definitions of Interaction Design, but the one that I prefer is:

Interaction Design is the design of any interactive experience.

In today's world, Interaction Design is concerned with the creation of meaningful experiences between us (humans) and objects. It is a good way to explore the creation of beautiful—and maybe even controversial—experiences between us and technology. Interaction Design encourages design through an iterative process based on prototypes of ever-increasing fidelity. This approach—also part of some types of "conventional" design—can be extended to include prototyping with technology; in particular, prototyping with electronics.

The specific field of Interaction Design involved with Arduino is Physical Computing (or Physical Interaction Design).

What Is Physical Computing?

Physical Computing uses electronics to prototype new materials for designers and artists.

It involves the design of interactive objects that can communicate with humans using sensors and actuators controlled by a behaviour implemented as software running inside a microcontroller (a small computer on a single chip).

In the past, using electronics meant having to deal with engineers all the time, and building circuits one small component at the time; these issues kept creative people from playing around with the medium directly. Most of the tools were meant for engineers and required extensive knowledge. In recent years, microcontrollers have become cheaper and easier to use, allowing the creation of better tools.

The progress that we have made with Arduino is to bring these tools one step closer to the novice, allowing people to start building stuff after only two or three days of a workshop.

With Arduino, a designer or artist can get to know the basics of electronics and sensors very quickly and can start building prototypes with very little investment.

2/The Arduino Way

The Arduino philosophy is based on making designs rather than talking about them. It is a constant search for faster and more powerful ways to build better prototypes. We have explored many prototyping techniques and developed ways of thinking with our hands.

Classic engineering relies on a strict process for getting from A to B; the Arduino Way delights in the possibility of getting lost on the way and finding C instead.

This is the tinkering process that we are so fond of—playing with the medium in an open-ended way and finding the unexpected. In this search for ways to build better prototypes, we also selected a number of software packages that enable the process of constant manipulation of the software and hardware medium.

The next few sections present some philosophies, events, and pioneers that have inspired the Arduino Way.

Prototyping

Prototyping is at the heart of the Arduino Way: we make things and build objects that interact with other objects, people, and networks. We strive to find a simpler and faster way to prototype in the cheapest possible way.

A lot of beginners approaching electronics for the first time think that they have to learn how to build everything from scratch. This is a waste of energy: what you want is to be able to confirm that something's working very quickly so that you can motivate yourself to take the next step or maybe even motivate somebody else to give you a lot of cash to do it.

This is why we developed "opportunistic prototyping": why spend time and energy building from scratch, a process that requires time and in-depth technical knowledge, when we can take ready-made devices and hack them in order to exploit the hard work done by large companies and good engineers?

Our hero is James Dyson, who made 5127 prototypes of his vacuum cleaner before he was satisfied that he'd gotten it right (**www.international. dyson.com/jd/1947.asp**).

Tinkering

We believe that it is essential to play with technology, exploring different possibilities directly on hardware and software—sometimes without a very defined goal.

Reusing existing technology is one of the best ways of tinkering. Getting cheap toys or old discarded equipment and hacking them to make them do something new is one of the best ways to get to great results.

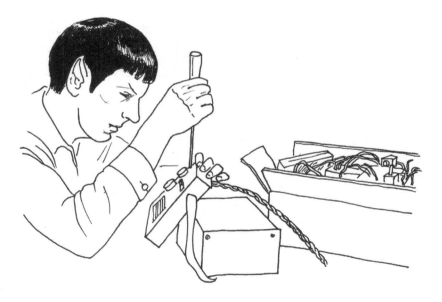

Patching

I have always been fascinated by modularity and the ability to build complex systems by connecting together simple devices. This process is very well represented by Robert Moog and his analogue synthesizers. Musicians constructed sounds, trying endless combinations by "patching together" different modules with cables. This approach made the synthesizer look like an old telephone switch, but combined with the numerous knobs, that was the perfect platform for tinkering with sound and innovating music. Moog described it as a process between "witnessing and discovering". I'm sure most musicians at first didn't know what all those hundreds of knobs did, but they tried and tried, refining their own style with no interruptions in the flow.

Reducing the number of interruptions to the flow is very important for creativity—the more seamless the process, the more tinkering happens.

This technique has been translated into the world of software by "visual programming" environments like Max, Pure Data, or VVVV. These tools can be visualised as "boxes" for the different functionalities that they provide, letting the user build "patches" by connecting these boxes together. These environments let the user experiment with programming without the constant interruption typical of the usual cycle: "type program, compile, damn—there is an error, fix error, compile, run". If you are more visually minded, I recommend that you try them out.

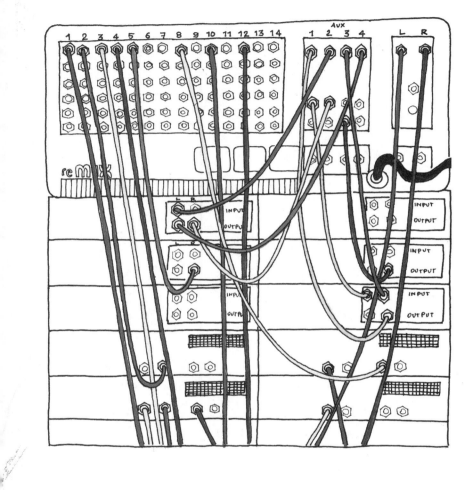

Circuit Bending

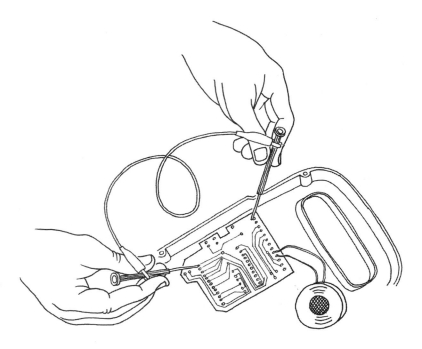

Circuit bending is one of the most interesting forms of tinkering. It's the creative short-circuiting of low-voltage, battery-powered electronic audio devices such as guitar effect pedals, children's toys, and small synthesizers to create new musical instruments and sound generators. The heart of this process is the "art of chance". It began in 1966 when Reed Ghazala, by chance, shorted-out a toy amplifier against a metal object in his desk drawer, resulting in a stream of unusual sounds. What I like about circuit benders is their ability to create the wildest devices by tinkering away with technology without necessarily understanding what they are doing on the theoretical side.

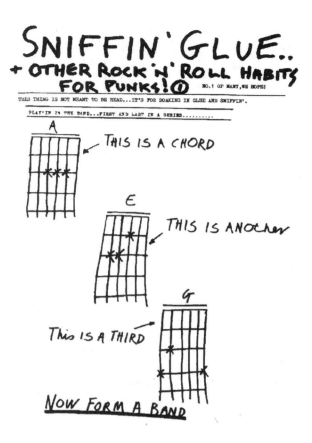

It's a bit like the *Sniffin' Glue* fanzine shown here: during the punk era, knowing three chords on a guitar was enough to start a band. Don't let the experts in one field tell you that you'll never be one of them. Ignore them and surprise them.

Keyboard Hacks

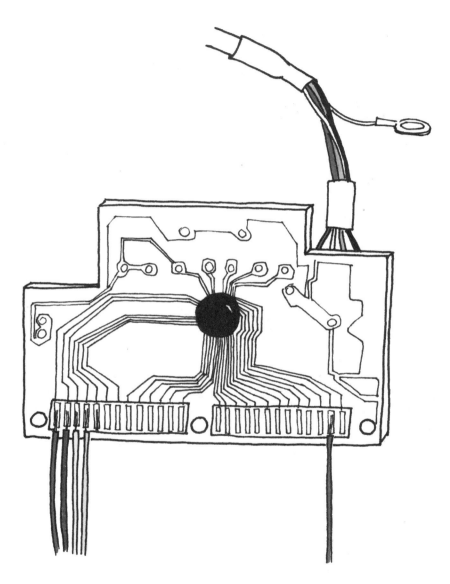

Computer keyboards are still the main way to interact with a computer after more than 60 years. Alex Pentland, academic head of the MIT Media Laboratory, once remarked: "Excuse the expression, but men's urinals are smarter than computers. Computers are isolated from what's around them."[1]

As tinkerers, we can implement new ways to interact with software by replacing the keys with devices that are able to sense the environment. Taking apart a computer keyboard reveals a very simple (and cheap) device. The heart of it is a small board. It's normally a smelly green or brown circuit with two sets of contacts going to two plastic layers that hold the connections between the different keys. If you remove the circuit and use a wire to bridge two contacts, you'll see a letter appear on the computer screen. If you go out and buy a motion-sensing detector and connect this to your keyboard, you'll see a key being pressed every time somebody walks in front of the computer. Map this to your favourite software, and you have made your computer as smart as a urinal. Learning about keyboard hacking is a key building block of prototyping and Physical Computing.

[1]Quoted in Sara Reese Hedberg, "MIT Media Lab's quest for perceptive computers," Intelligent Systems and Their Applications, IEEE, Jul/Aug 1998.

We Love Junk!

People throw away a lot of technology these days: old printers, computers, weird office machines, technical equipment, and even a lot of military
stuff. There has always been a big market for this surplus technology,
especially among young and/or poorer hackers and those who are just
starting out. This market become evident in Ivrea, where we developed
Arduino. The city used to be the headquarters of the Olivetti company.
They had been making computers since the 1960s; in the mid 1990s, they
threw everything away in junkyards in the area. These are full of computer parts, electronic components, and weird devices of all kinds. We
spent countless hours there, buying all sorts of contraptions for very little
money and hacking into our prototypes. When you can buy a thousand
loudspeakers for very little money, you're bound to come up with some
idea in the end. Accumulate junk and go through it before starting to build
something from scratch.

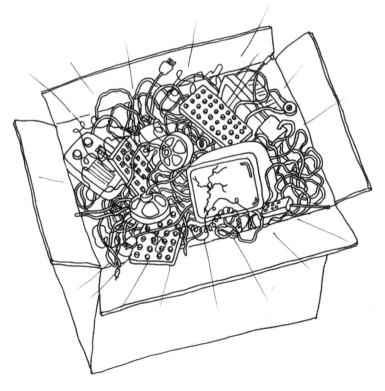

Hacking Toys

Toys are a fantastic source of cheap technology to hack and reuse, as evidenced by the practise of circuit bending mentioned earlier. With the current influx of thousands of very cheap high-tech toys from China, you can build quick ideas with a few noisy cats and a couple of light swords. I have been doing this for a few years to get my students to understand that technology is not scary or difficult to approach. One of my favourite resources is the booklet "Low Tech Sensors and Actuators" by Usman Haque and Adam Somlai-Fischer (**lowtech.propositions.org.uk**). I think that they have perfectly described this technique in that handbook, and I have been using it ever since.

Collaboration

Collaboration between users is one of they key principles in the Arduino world—through the forum at www.arduino.cc, people from different parts of the world help each other learn about the platform. The Arduino team encourages people to collaborate at a local level as well by helping them set up users' groups in every city they visit. We also set up a Wiki called "Playground" (www.arduino.cc/playground) where users document their findings. It's so amazing to see how much knowledge these people pour out on the Web for everybody to use. This culture of sharing and helping each other is one of the things that I'm most proud of in regard to Arduino.

3/The Arduino Platform

Arduino is composed of two major parts: the Arduino board, which is the piece of hardware you work on when you build your objects; and the Arduino IDE, the piece of software you run on your computer. You use the IDE to create a sketch (a little computer program) that you upload to the Arduino board. The sketch tells the board what to do.

Not too long ago, working on hardware meant building circuits from scratch, using hundreds of different components with strange names like resistor, capacitor, inductor, transistor, and so on.

Every circuit was "wired" to do one specific application, and making changes required you to cut wires, solder connections, and more.

With the appearance of digital technologies and microprocessors, these functions, which were once implemented with wires, were replaced by software programs.

Software is easier to modify than hardware. With a few keypresses, you can radically change the logic of a device and try two or three versions in the same amount of time that it would take you to solder a couple of resistors.

The Arduino Hardware

The Arduino board is a small microcontroller board, which is a small circuit (the board) that contains a whole computer on a small chip (the microcontroller). This computer is at least a thousand times less powerful than the MacBook I'm using to write this, but it's a lot cheaper and very useful to build interesting devices. Look at the Arduino board: you'll see a black chip with 28 "legs"—that chip is the ATmega168, the heart of your board.

We (the Arduino team) have placed on this board all the components that are required for this microcontroller to work properly and to communicate with your computer. There are many versions of this board; the one we'll use throughout this book is the Arduino Duemilanove, which is the simplest one to use and the best one for learning on. However, these instructions apply to earlier versions of the board, including the more recent Arduino Diecimila and the older Arduino NG. Figure 3-1 shows the Arduino Duemilanove; Figure 3-2 shows the Arduino NG.

In those illustrations, you see the Arduino board. At first, all those connectors might be a little confusing. Here is an explanation of what every element of the board does:

14 Digital IO pins (pins 0–13)
These can be inputs or outputs, which is specified by the sketch you create in the IDE.

6 Analogue In pins (pins 0–5)
These dedicated analogue input pins take analogue values (i.e., voltage readings from a sensor) and convert them into a number between 0 and 1023.

6 Analogue Out pins (pins 3, 5, 6, 9, 10, and 11)
These are actually six of the digital pins that can be reprogrammed for analogue output using the sketch you create in the IDE.

The board can be powered from your computer's USB port, most USB chargers, or an AC adapter (9 volts recommended, 2.1mm barrel tip, center positive). If there is no power supply plugged into the power socket, the power will come from the USB board, but as soon as you plug a power supply, the board will automatically use it.

NOTE: If you are using the older Arduino-NG or Arduino Diecimila, you will need to set the power selection jumper (labelled PWR_SEL on the board) to specify EXT (external) or USB power. This jumper can be found between the plug for the AC adapter and the USB port.

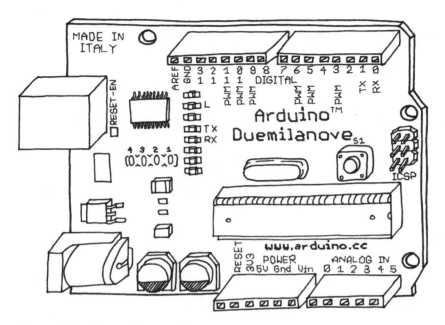

Figure 3-1. The Arduino Duemilanove

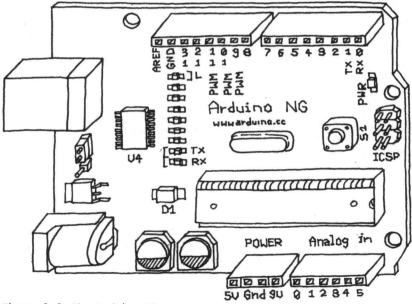

Figure 3-2. The Arduino NG

The Software (IDE)

The IDE (Integrated Development Environment) is a special program running on your computer that allows you to write sketches for the Arduino board in a simple language modeled after the Processing (www.processing.org) language. The magic happens when you press the button that uploads the sketch to the board: the code that you have written is translated into the C language (which is generally quite hard for a beginner to use), and is passed to the avr-gcc compiler, an important piece of open source software that makes the final translation into the language understood by the microcontroller. This last step is quite important, because it's where Arduino makes your life simple by hiding away as much as possible of the complexities of programming microcontrollers.

The programming cycle on Arduino is basically as follows:

» Plug your board into a USB port on your computer.

» Write a sketch that will bring the board to life.

» Upload this sketch to the board through the USB connection and wait a couple of seconds for the board to restart.

» The board executes the sketch that you wrote.

NOTE: Installing Arduino on Linux is somewhat complicated at the time of this writing. See www.arduino.cc/playground/Learning/Linux for complete instructions.

Installing Arduino on Your Computer

To program the Arduino board, you must first download the development environment (the IDE) from here: www.arduino.cc/en/Main/Software. Choose the right version for your operating system.

Download the file and double-click on it to uncompress it; this will create a folder named *arduino-[version]*, such as *arduino-0012*. Drag that folder to wherever you would like it to be: your desktop, your */Applications* folder (on a Mac), or your *C:\Program Files* folder (on Windows). Now whenever you want to run the Arduino IDE, you'll open up the *arduino* folder, and double-click the Arduino icon. Don't do this just yet, though; there is one more step to perform.

NOTE: If you have any trouble running the Arduino IDE, see Chapter 7, Troubleshooting.

Now you must install the drivers that allow your computer to talk to your board through the USB port.

Installing Drivers: Macintosh

Look for the *Drivers* folder inside the *arduino-0012* folder and double-click the file called *FTDIUSBSerialDriver_x_x_x.dmg* (*x_x_x* will be replaced with the version number of the driver, for example *FTDIUSBSerialDriver_v2_2_9_Intel.dmg*). Double-click the *.dmg* file to mount it.

NOTE: If you are using an Intel-based Mac, such as a MacBook, Mac-Book Pro, MacBook Air, Mac Pro, or Intel-based Mac Mini or iMac, be sure to install the driver with "Intel" in its name, as in *FTDIUSBSerial-Driver_v2_2_9_Intel.dmg*. If you aren't using an Intel-based Mac, install the one without "Intel" in its name.

Next, install the software from the FTDIUSBSerialDriver package by double-clicking on it. Follow the instructions provided by the installer and type the password of an administrative user if asked. At the end of this process, restart your machine to make sure that the drivers are properly loaded. Now plug the board into your computer. The PWR light on the board should come on and the yellow LED labelled "L" should start blinking. If not, see Chapter 7, Troubleshooting.

Installing Drivers: Windows

Plug the Arduino board into the computer; when the Found New Hardware Wizard window comes up, Windows will first try to find the driver on the Windows Update site.

Windows XP will ask you whether to check Windows Update; if you don't want to use Windows Update, select the "No, not at this time" option and click Next.

On the next screen, choose "Install from a list or specific location" and click Next.

Check the box labeled "Include this location in the search", click Browse, select the folder where you installed Arduino, and select the *Drivers\FTDI USB Drivers* folder as the location. Click OK, and Next.

Windows Vista will first attempt to find the driver on Windows Update; if that fails, you can instruct it to look in the *Drivers\FTDI USB Drivers* folder.

You'll go through this procedure twice, because the computer first installs the low-level driver, then installs a piece of code that makes the board look like a serial port to the computer.

Once the drivers are installed, you can launch the Arduino IDE and start using Arduino.

Next, you must figure out which serial port is assigned to your Arduino board—you'll need that information to program it later. The instructions for getting this information are in the following sections.

Port Identification: Macintosh

From the Tools menu in the Arduino IDE, select "Serial Port" and select the port that begins with /dev/cu.usbserial-; this is the name that your computer uses to refer to the Arduino board. Figure 3-3 shows the list of ports.

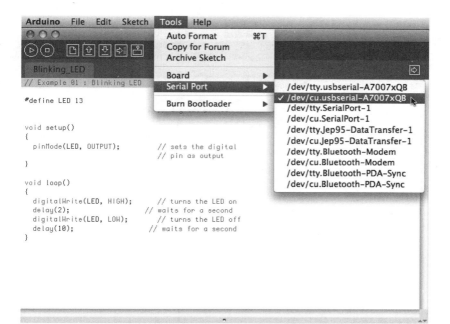

Figure 3-3.
The Arduino IDE's list of serial ports

Port Identification: Windows

On Windows, the process is a bit more complicated—at least at the beginning. Open the Device Manager by clicking the Start menu, right-clicking on Computer (Vista) or My Computer (XP), and choosing Properties. On Windows XP, click Hardware and choose Device Manager. On Vista, click Device Manager (it appears in the list of tasks on the left of the window).

Look for the Arduino device in the list under "Ports (COM & LPT)". The Arduino will appear as a USB Serial Port and will have a name like COM3, as shown in Figure 3-4.

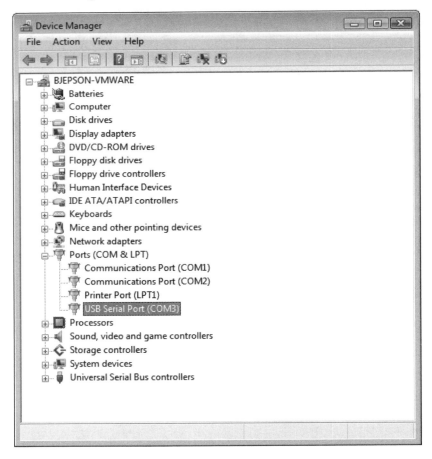

Figure 3-4.
The Windows Device Manager showing all available serial ports

---

NOTE: On some Windows machines, the COM port has a number greater than 9; this numbering creates some problems when Arduino is trying to communicate with it. See Chapter 7, Troubleshooting for help on this problem.

Once you've figured out the COM port assignment, you can select that port from the Tools > Serial Port menu in the Arduino IDE.

Now the Arduino development environment can talk to the Arduino board and program it.

4/Really Getting Started with Arduino

Now you'll learn how to build and program an interactive device.

Anatomy of an Interactive Device

All of the objects we will build using Arduino follow a very simple pattern that we call the "Interactive Device". The Interactive Device is an electronic circuit that is able to sense the environment using sensors (electronic components that convert real-world measurements into electrical signals). The device processes the information it gets from the sensors with behaviour that's implemented as software. The device will then be able to interact with the world using actuators, electronic components that can convert an electric signal into a physical action.

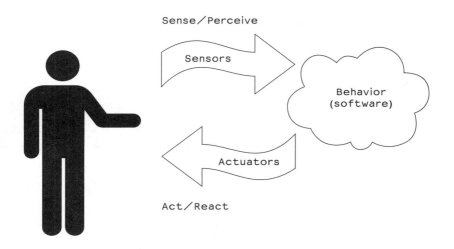

Figure 4-1.
The interactive device

Sensors and Actuators

Sensors and actuators are electronic components that allow a piece of electronics to interact with the world.

As the microcontroller is a very simple computer, it can process only electric signals (a bit like the electric pulses that are sent between neurons in our brains). For it to sense light, temperature, or other physical quantities, it needs something that can convert them into electricity. In our body, for example, the eye converts light into signals that get sent to the brain using nerves. In electronics, we can use a simple device called a light-dependent resistor (an LDR or photoresistor) that can measure the amount of light that hits it and report it as a signal that can be understood by the micro-controller.

Once the sensors have been read, the device has the information needed to decide how to react. The decision-making process is handled by the microcontroller, and the reaction is performed by actuators. In our bodies, for example, muscles receive electric signals from the brain and convert them into a movement. In the electronic world, these functions could be performed by a light or an electric motor.

In the following sections, you will learn how to read sensors of different types and control different kinds of actuators.

Blinking an LED

The LED blinking sketch is the first program that you should run to test whether your Arduino board is working and is configured correctly. It is also usually the very first programming exercise someone does when learning to program a microcontroller. A light-emitting diode (LED) is a small electronic component that's a bit like a light bulb, but is more efficient and requires lower voltages to operate.

Your Arduino board comes with an LED preinstalled. It's marked "L". You can also add your own LED—connect it as shown in Figure 4-2.

K indicates the cathode (negative), or shorter lead; A indicates the anode (positive), or longer lead.

Once the LED is connected, you need to tell Arduino what to do. This is done through code, that is, a list of instructions that we give the micro-controller to make it do what we want.

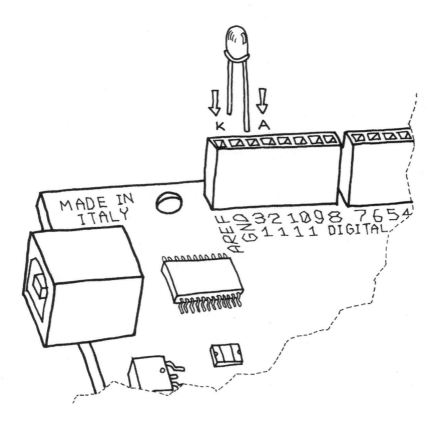

Figure 4-2.
Connecting an LED to Arduino

On your computer, go open the folder where you copied the Arduino IDE. Double-click the Arduino icon to start it. Select File > New and you'll be asked to choose a sketch folder name: this is where your Arduino sketch will be stored. Name it *Blinking_LED* and click OK. Then, type the following text (Example 01) into the Arduino sketch editor (the main window of the Arduino IDE). You can also download it from **www.makezine.com/ getstartedarduino**. It should appear as shown in Figure 4-3.

```
// Example 01 : Blinking LED

#define LED 13    // LED connected to
                  // digital pin 13

void setup()
{
  pinMode(LED, OUTPUT);    // sets the digital
                           // pin as output
}

void loop()
{
  digitalWrite(LED, HIGH);  // turns the LED on
  delay(1000);              // waits for a second
  digitalWrite(LED, LOW);   // turns the LED off
  delay(1000);              // waits for a second
}
```

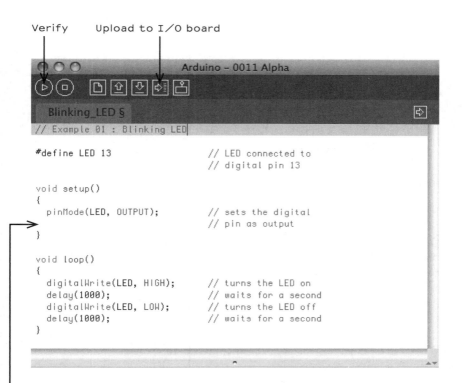

Verify   Upload to I/O board

```
// Example 01 : Blinking LED

#define LED 13                    // LED connected to
                                  // digital pin 13

void setup()
{
  pinMode(LED, OUTPUT);           // sets the digital
                                  // pin as output
}

void loop()
{
  digitalWrite(LED, HIGH);        // turns the LED on
  delay(1000);                    // waits for a second
  digitalWrite(LED, LOW);         // turns the LED off
  delay(1000);                    // waits for a second
}
```

Your sketch goes here

Figure 4-3.
The Arduino IDE with your first sketch loaded

Now that the code is in your IDE, you need to verify that it is correct. Press the "Verify" button (Figure 4-3 shows its location); if everything is correct, you'll see the message "Done compiling" appear at the bottom of the Arduino IDE. This message means that the Arduino IDE has translated your sketch into an executable program that can be run by the board, a bit like an .exe file in Windows or an .app file on a Mac.

At this point, you can upload it into the board: press the Upload to I/O Board button (see Figure 4-3). This will reset the board, forcing it to stop what it's doing and listen for instructions coming from the USB port. The Arduino IDE sends the current sketch to the board, which will store it in its memory and eventually run it.

You will see a few messages appear in the black area at the bottom of the window, and just above that area, you'll see the message "Done uploading" appear to let you know the process has completed correctly. There are two LEDs, marked RX and TX, on the board; these flash every time a byte is sent or received by the board. During the upload process, they keep flickering.

If you don't see the LEDs flicker, or if you get an error message instead of "Done uploading", then there is a communication problem between your computer and Arduino. Make sure you've selected the right serial port (see Chapter 3) in the Tools > Serial Port menu. Also, check the Tools > Board menu to confirm that the correct model of Arduino is selected there.

If you are still having problems, check Chapter 7, Troubleshooting.

Once the code is in your Arduino board, it will stay there until you put another sketch on it. The sketch will survive if the board is reset or turned off, a bit like the data on your computer's hard drive.

Assuming that the sketch has been uploaded correctly, you will see the LED "L" turn on for a second and then turn off for a second. If you installed a separate LED as shown back in Figure 4-2, that LED will blink, too. What you have just written and ran is a "computer program", or sketch, as Arduino programs are called. Arduino, as I've mentioned before, is a small computer, and it can be programmed to do what you want. This is done using a programming language to type a series of instructions in the Arduino IDE, which turns it into an executable for your Arduino board.

I'll next show you how to understand the sketch. First of all, the Arduino executes the code from top to bottom, so the first line at the top is the first one read; then it moves down, a bit like how the playhead of a video player like QuickTime Player or Windows Media Player moves from left to right showing where in the movie you are.

Pass Me the Parmesan

Notice the presence of curly brackets, which are used to group together lines of code. These are particularly useful when you want to give a name to a group of instructions. If you're at dinner and you ask somebody, "Please pass me the Parmesan cheese," this kicks off a series of actions that are summarised by the small phrase that you just said. As we're humans, it all comes naturally, but all the individual tiny actions required to do this must be spelled out to the Arduino, because it's not as powerful

as our brain. So to group together a number of instructions, you stick a {
before your code and an } after.

You can see that there are two blocks of code that are defined in this way
here. Before each one of them there is a strange command:

```
void setup()
```

This line gives a name to a block of code. If you were to write a list of
instructions that teach Arduino how to pass the Parmesan, you would
write *void passTheParmesan()* at the beginning of a block, and this block
would become an instruction that you can call from anywhere in the
Arduino code. These blocks are called **functions**. If after this, you write
passTheParmesan() anywhere in your code, Arduino will execute those
instructions and continue where it left off.

Arduino Is Not for Quitters

Arduino expects two functions to exists—one called *setup()* and one
called *loop()*.

setup() is where you put all the code that you want to execute once at the
beginning of your program and *loop()* contains the core of your program,
which is executed over and over again. This is done because Arduino is
not like your regular computer—it cannot run multiple programs at the
same time and programs can't quit. When you power up the board, the
code runs; when you want to stop, you just turn it off.

Real Tinkerers Write Comments

Any text beginning with *//* is ignored by Arduino. These lines are comments,
which are notes that you leave in the program for yourself, so that you can
remember what you did when you wrote it, or for somebody else, so that
they can understand your code.

It is very common (I know this because I do it all the time) to write a piece
of code, upload it onto the board, and say "Okay—I'm never going to have
to touch this sucker again!" only to realise six months later that you need
to update the code or fix a bug. At this point, you open up the program,
and if you haven't included any comments in the original program, you'll
think, "Wow—what a mess! Where do I start?" As we move along, you'll
see some tricks for how to make your programs more readable and easier
to maintain.

The Code, Step by Step

At first, you might consider this kind of explanation too unnecessary, a bit like when I was in school and I had to study Dante's *Divina Commedia* (every Italian student has to go through that, as well as another book called *I promessi sposi*, or *The Betrothed*—oh, the nightmares). For each line of the poems, there were a hundred lines of commentary! However, the explanation will be much more useful here as you move on to writing your own programs.

```
// Example 01 : Blinking LED
```
A comment is a useful way for us to write little notes. The preceding title comment just reminds us that this program, Example 01, blinks an LED.

```
#define LED 13   // LED connected to
                 // digital pin 13
```
#define is like an automatic search and replace for your code; in this case, it's telling Arduino to write the number 13 every time the word *LED* appears. The replacement is the first thing done when you click Verify or Upload to I/O Board (you never see the results of the replacement as it's done behind the scenes). We are using this command to specify that the LED we're blinking is connected to the Arduino pin 13.

```
void setup()
```
This line tells Arduino that the next block of code will be called *setup()*.

```
{
```
With this opening curly bracket, a block of code begins.

```
pinMode(LED, OUTPUT); // sets the digital
                      // pin as output
```
Finally, a really interesting instruction. *pinMode* tells Arduino how to configure a certain pin. Digital pins can be used either as INPUT or OUTPUT. In this case, we need an output pin to control our LED, so we place the number of the pin and its mode inside the parentheses. *pinMode* is a function, and the words (or numbers) specified inside the parentheses are **arguments**. INPUT and OUTPUT are constants in the Arduino language. (Like variables, constants are assigned values, except that constant values are predefined and never change.)

```
}
```
This closing curly bracket signifies the end of the *setup()* function.

```
void loop()
{
```

loop() is where you specify the main behaviour of your interactive device. It will be repeated over and over again until you switch the board off.

```
digitalWrite(LED, HIGH);   // turns the LED on
```
As the comment says, *digitalWrite()* is able to turn on (or off) any pin that has been configured as an OUTPUT. The first argument (in this case, *LED*) specifies which pin should be turned on or off (remember that *LED* is a constant value that refers to pin 13, so this is the pin that's switched). The second argument can turn the pin on (HIGH) or off (LOW).

Imagine that every output pin is a tiny power socket, like the ones you have on the walls of your apartment. European ones are 230 V, American ones are 110 V, and Arduino works at a more modest 5 V. The magic here is when software becomes hardware. When you write *digitalWrite(LED, HIGH)*, it turns the output pin to 5 V, and if you connect an LED, it will light up. So at this point in your code, an instruction in software makes something happen in the physical world by controlling the flow of electricity to the pin. Turning on and off the pin at will now let us translate these into something more visible for a human being; the LED is our actuator.

```
delay(1000);      // waits for a second
```
Arduino has a very basic structure. Therefore, if you want things to happen with a certain regularity, you tell it to sit quietly and do nothing until it is time to go to the next step. *delay()* basically makes the processor sit there and do nothing for the amount of milliseconds that you pass as an argument. Milliseconds are thousandths of seconds; therefore, 1000 milliseconds equals 1 second. So the LED stays on for one second here.

```
digitalWrite(LED, LOW);      // turns the LED off
```
This instruction now turns off the LED that we previously turned on. Why do we use HIGH and LOW? Well, it's an old convention in digital electronics. HIGH means that the pin is on, and in the case of Arduino, it will be set at 5 V. LOW means 0 V. You can also replace these arguments mentally with ON and OFF.

```
delay(1000); // waits for a second
```
Here, we delay for another second. The LED will be off for one second.

```
}
```
This closing curly bracket marks end of the loop function.

To sum up, this program does this:

» Turns pin 13 into an output (just once at the beginning)

» Enters a loop

» Switches on the LED connected to pin 13

» Waits for a second

» Switches off the LED connected to pin 13

» Waits for a second

» Goes back to beginning of the loop

I hope that wasn't too painful. You'll learn more about how to program as you go through the later examples.

Before we move on to the next section, I want you to play with the code. For example, reduce the amount of delay, using different numbers for the on and off pulses so that you can see different blinking patterns. In particular, you should see what happens when you make the delays very small, but use different delays for on and off . . . there is a moment when something strange happens; this "something" will be very useful when you learn about pulse-width modulation later in this book.

What We Will Be Building

I have always been fascinated by light and the ability to control different light sources through technology. I have been lucky enough to work on some interesting projects that involve controlling light and making it interact with people. Arduino is really good at this. Throughout this book, we will be working on how to design "interactive lamps", using Arduino as a way to learn the basics of how interactive devices are built.

In the next section, I'll try to explain the basics of electricity in a way that would bore an engineer, but won't scare a new Arduino programmer.

What Is Electricity?

If you have done any plumbing at home, electronics won't be a problem for you to understand. To understand how electricity and electric circuits work, the best way is to use something called the "water analogy". Let's take a simple device, like the battery-powered portable fan shown in Figure 4-4.

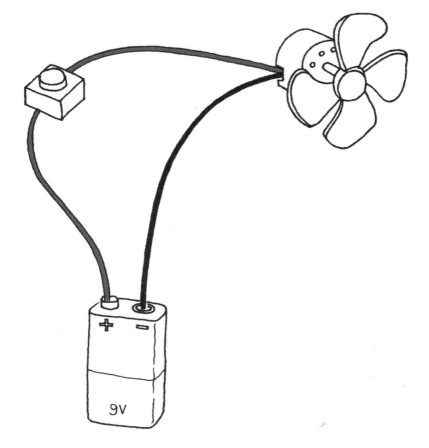

Figure 4-4.
A portable fan

If you take a fan apart, you will see that it contains a small battery, a couple of wires, and an electric motor, and that one of the wires going to the motor is interrupted by a switch. If you have a fresh battery in it and turn the switch on, the motor will start to spin, providing the necessary chill. How does this work? Well, imagine that the battery is both a water reservoir and a pump, the switch is a tap, and the motor is one of those wheels that you see in watermills. When you open the tap, water flows from the pump and pushes the wheel into motion.

In this simple hydraulic system, shown in Figure 4-5, two factors are important: the pressure of the water (this is determined by the power of pump) and the amount of water that will flow in the pipes (this depends on the size of the pipes and the resistance that the wheel will provide to the stream of water hitting it).

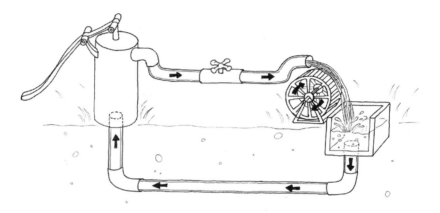

Figure 4-5.
A hydraulic system

You'll quickly realise that if you want the wheel to spin faster, you need to increase the size of the pipes (but this works only up to a point) and increase the pressure that the pump can achieve. Increasing the size of the pipes allows a greater flow of water to go through them; by making them bigger, we have effectively reduced the pipes' resistance to the flow of water. This approach works up to a certain point, at which the wheel won't spin any faster, because the pressure of the water is not strong enough. When we reach this point, we need the pump to be stronger. This method of speeding up the watermill can go on until the point when the wheel falls apart because the water flow is too strong for it and it is destroyed. Another thing you will notice is that as the wheel spins, the axle will heat up a little bit, because no matter how well we have mounted the wheel, the friction between the axle and the holes in which it is mounted in will generate heat. It is important to understand that in a system like this, not all the energy you pump into the system will be converted into movement; some will be lost in a number of inefficiencies and will generally show up as heat emanating from some parts of the system.

So what are the important parts of the system? The pressure produced by the pump is one; the resistance that the pipes and wheel offer to the flow of water, and the actual flow of water (let's say that this is represented by the number of litres of water that flow in one second) are the others. Electricity works a bit like water. You have a kind of pump (any source of electricity, like a battery or a wall plug) that pushes electric charges (imagine them as "drops" of electricity) down pipes, which are represented by the wires—some devices are able to use these to produce heat (your grandma's thermal blanket), light (your bedroom lamp), sound (your stereo), movement (your fan), and much more.

So when you read that a battery's voltage is 9 V, think of this voltage like the water pressure that can potentially be produced by this little "pump". Voltage is measured in volts, named after Alessandro Volta, the inventor of the first battery.

Just as water pressure has an electric equivalent, the flow rate of water does, too. This is called current, and is measured in amperes (after André-Marie Ampère, electromagnetism pioneer). The relationship between voltage and current can be illustrated by returning to the water wheel: a higher voltage (pressure) lets you spin a wheel faster; a higher flow rate (current) lets you spin a larger wheel.

Finally, the resistance opposing the flow of current over any path that it travels is called—you guessed it—resistance, and is measured in ohms (after the German physicist Georg Ohm). Herr Ohm was also responsible for formulating the most important law in electricity—and the only formula that you really need to remember. He was able to demonstrate that in a circuit the voltage, the current, and the resistance are all related to each other, and in particular that the resistance of a circuit determines the amount of current that will flow through it, given a certain voltage supply.

It's very intuitive, if you think about it. Take a 9 V battery and plug it into a simple circuit. While measuring current, you will find that the more resistors you add to the circuit, the less current will travel through it. Going back to the analogy of water flowing in pipes, given a certain pump, if I install a valve (which we can relate to a **variable resistor** in electricity), the more

I close the valve—increasing resistance to water flow—the less water will flow through the pipes. Ohm summarised his law in these formulae:

```
R (resistance) = V (voltage) / I (current)
V = R * I
I = V / R
```

This is the only rule that you really have to memorise and learn to use, because in most of your work, this is the only one that you will really need.

Using a Pushbutton to Control the LED

Blinking an LED was easy, but I don't think you would stay sane if your desk lamp were to continuously blink while you were trying to read a book. Therefore, you need to learn how to control it. In our previous example, the LED was our actuator, and our Arduino was controlling it. What is missing to complete the picture is a sensor.

In this case, we're going to use the simplest form of sensor available: a pushbutton.

If you were to take apart a pushbutton, you would see that it is a very simple device: two bits of metal kept apart by a spring, and a plastic cap that when pressed brings the two bits of metal into contact. When the bits of metal are apart, there is no circulation of current in the pushbutton (a bit like when a water valve is closed); when we press it, we make a connection.

To monitor the state of a switch, there's a new Arduino instruction that you're going to learn: the *digitalRead()* function.

digitalRead() checks to see whether there is any voltage applied to the pin that you specify between parentheses, and returns a value of HIGH or LOW, depending on its findings. The other instructions that we've used so far haven't returned any information—they just executed what we asked them to do. But that kind of function is a bit limited, because it will force us to stick with very predictable sequences of instructions, with no input from the outside world. With *digitalRead()*, we can "ask a question" of Arduino and receive an answer that can be stored in memory somewhere and used to make decisions immediately or later.

Build the circuit shown in Figure 4-6. To build this, you'll need to obtain some parts (these will come in handy as you work on other projects as well):

» Solderless breadboard: RadioShack (www.radioshack.com) part number 276-002, Maker Shed (www.makershed.com) part number MKKN3. Appendix A is an introduction to the solderless breadboard.

» Pre-cut jumper wire kit: RadioShack 276-173, Maker Shed MKKN4

» One 10K Ohm resistor: RadioShack 271-1335 (5-pack), SparkFun (www.sparkfun.com) COM-08374

» Momentary tactile pushbutton switch: SparkFun COM-00097

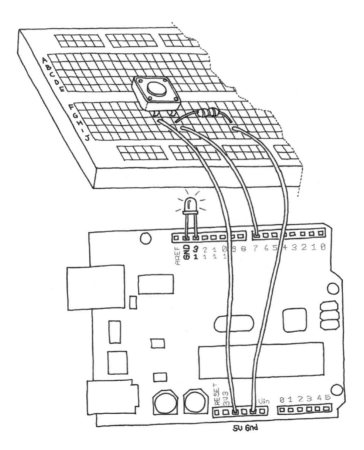

Figure 4-6.
Hooking up a pushbutton

Let's have a look at the code that we'll be using to control the LED with our
pushbutton:

```
// Example 02: Turn on LED while the button is pressed

#define LED 13    // the pin for the LED
#define BUTTON 7  // the input pin where the
                  // pushbutton is connected
int val = 0;      // val will be used to store the state
                  // of the input pin

void setup() {
  pinMode(LED, OUTPUT);    // tell Arduino LED is an output
  pinMode(BUTTON, INPUT);  // and BUTTON is an input
}

void loop(){
  val = digitalRead(BUTTON); // read input value and store it

  // check whether the input is HIGH (button pressed)
  if (val == HIGH) {
    digitalWrite(LED, HIGH); // turn LED ON
  } else {
    digitalWrite(LED, LOW);
  }
}
```

In Arduino, select File > New (if you have another sketch open, you may
want to save it first). When Arduino asks you to name your new sketch
folder, type PushButtonControl. Type the Example 02 code into Arduino
(or download it from www.makezine.com/getstartedarduino and paste
it into the Arduino IDE). If everything is correct, the LED will light up when
you press the button.

How Does This Work?

I have introduced two new concepts with this example program: functions
that return the result of their work and the *if* statement.

The *if* statement is possibly the most important instruction in a programming language, because it allows the computer (and remember, the Arduino is a small computer) to make decisions. After the *if* keyword, you have to write a "question" inside parentheses, and if the "answer", or result, is true, the first block of code will be executed; otherwise, the block of code after else will be executed. Notice that I have used the == symbol instead of =. The former is used when two entities are compared, and returns TRUE or FALSE; the latter assigns a value to a variable. Make sure that you use the correct one, because it is very easy to make that mistake and use just =, in which case your program will never work. I know, because after 25 years of programming, I still make that mistake.

Holding your finger on the button for as long as you need light is not practical. Although it would make you think about how much energy you're wasting when you walk away from a lamp that you left on, we need to figure out how to make the on button "stick".

One Circuit, A Thousand Behaviours

The great advantage of digital, programmable electronics over classic electronics now becomes evident: I will show you how to implement many different "behaviours" using the same electronic circuit as in the previous section, just by changing the software.

As I've mentioned before, it's not very practical to have to hold your finger on the button to have the light on. We therefore must implement some form of "memory", in the form of a software mechanism that will remember when we have pressed the button and will keep the light on even after we have released it.

To do this, we're going to use what is called a **variable**. (We have used one already, but I haven't explained it.) A variable is a place in the Arduino memory where you can store data. Think of it like one of those sticky notes you use to remind yourself about something, such as a phone number: you take one, you write "Luisa 02 555 1212" on it, and you stick it to your computer monitor or your fridge. In the Arduino language, it's equally simple: you just decide what type of data you want to store (a number or some text, for example), give it a name, and when you want to, you can store the data or retrieve it. For example:

```
int val = 0;
```

int means that your variable will store an integer number, *val* is the name of the variable, and = *0* assigns it an initial value of zero.

A variable, as the name intimates, can be modified anywhere in your code, so that later on in your program, you could write:

```
val = 112;
```

which reassigns a new value, 112, to your variable.

NOTE: Have you noticed that in Arduino, every instruction, with one exception (#define), ends with a semicolon? This is done so that the compiler (the part of Arduino that turns your sketch into a program that the microcontroller can run) knows that your statement is finished and a new one is beginning. Remember to use it all the time, excluding any line that begins with #define. The #defines are replaced by the compiler before the code is translated into an Arduino executable.

In the following program, *val* is used to store the result of *digitalRead()*; whatever Arduino gets from the input ends up in the variable and will stay there until another line of code changes it. Notice that variables use a type of memory called RAM. It is quite fast, but when you turn off your board, all data stored in RAM is lost (which means that each variable is re-set to its initial value when the board is powered up again). Your programs themselves are stored in flash memory—this is the same type used by your mobile phone to store phone numbers—which retains its content even when the board is off.

Let's now use another variable to remember whether the LED has to stay on or off after we release the button. Example 03A is a first attempt at achieving that:

```
// Example 03A: Turn on LED when the button is pressed
// and keep it on after it is released


#define LED  13  // the pin for the LED
#define BUTTON 7 // the input pin where the
                 // pushbutton is connected
int val = 0;     // val will be used to store the state
                 // of the input pin
int state = 0;   // 0 = LED off while 1 = LED on

void setup() {
  pinMode(LED, OUTPUT);   // tell Arduino LED is an output
  pinMode(BUTTON, INPUT); // and BUTTON is an input
}


void loop() {
  val = digitalRead(BUTTON); // read input value and store it

  // check if the input is HIGH (button pressed)
  // and change the state
  if (val == HIGH) {
    state = 1 - state;
  }


  if (state == 1) {
    digitalWrite(LED, HIGH); // turn LED ON
  } else {
    digitalWrite(LED, LOW);
  }
}
```

Now go test this code. You will notice that it works . . . somewhat. You'll find that the light changes so rapidly that you can't reliably set it on or off with a button press.

Let's look at the interesting parts of the code: *state* is a variable that stores either 0 or 1 to remember whether the LED is on or off. After the button is released, we initialise it to 0 (LED off).

Later, we read the current state of the button, and if it's pressed (*val* == *HIGH*), we change state from 0 to 1, or vice versa. We do this using a small trick, as state can be only either 1 or 0. The trick I use involves a small mathematical expression based on the idea that 1 − 0 is 1 and 1 − 1 is 0:

```
state = 1 - state;
```

The line may not make much sense in mathematics, but it does in programming. The symbol = means "assign the result of what's after me to the variable name before me"—in this case, the new value of state is assigned the value of 1 minus the old value of state.

Later in the program, you can see that we use *state* to figure out whether the LED has to be on or off. As I mentioned, this leads to somewhat flaky results.

The results are flaky because of the way we read the button. Arduino is really fast; it executes its own internal instructions at a rate of 16 million per second—it could well be executing a few million lines of code per second. So this means that while your finger is pressing the button, Arduino might be reading the button's position a few thousand times and changing *state* accordingly. So the results end up being unpredictable; it might be off when you wanted it on, or vice versa. As even a broken clock is right twice a day, the program might show the correct behaviour every once in a while, but much of the time it will be wrong.

How do we fix this? Well, we need to detect the exact moment when the button is pressed—that is the only moment that we have to change state. The way I like to do it is to store the value of *val* before I read a new one; this allows me to compare the current position of the button with the previous one and change state only when the button becomes HIGH after being LOW.

Example 03B contains the code to do so:

```
// Example 03B: Turn on LED when the button is pressed
// and keep it on after it is released
// Now with a new and improved formula!


#define LED  13   // the pin for the LED
#define BUTTON 7  // the input pin where the
                  // pushbutton is connected
int val = 0;      // val will be used to store the state
                  // of the input pin
int old_val = 0;  // this variable stores the previous
                  // value of "val"
int state = 0;    // 0 = LED off and 1 = LED on

void setup() {
  pinMode(LED, OUTPUT);     // tell Arduino LED is an output
  pinMode(BUTTON, INPUT);   // and BUTTON is an input
}
void loop(){
  val = digitalRead(BUTTON); // read input value and store it
                             // yum, fresh

  // check if there was a transition
  if ((val == HIGH) && (old_val == LOW)){
    state = 1 - state;
  }

  old_val = val;  // val is now old, let's store it

  if (state == 1) {
    digitalWrite(LED, HIGH); // turn LED ON
  } else {
    digitalWrite(LED, LOW);
  }
}
```

Test it: we're almost there!

You may have noticed that this approach is not entirely perfect, due to another issue with mechanical switches. Pushbuttons are very simple devices: two bits of metal kept apart by a spring. When you press the button, the two contacts come together and electricity can flow. This sounds fine and simple, but in real life the connection is not that perfect, especially when the button is not completely pressed, and it generates some spurious signals called **bouncing**.

When the pushbutton is bouncing, the Arduino sees a very rapid sequence of on and off signals. There are many techniques developed to do de-bouncing, but in this simple piece of code I've noticed that it's usually enough to add a 10- to 50-millisecond delay when the code detects a transition.

Example 03C is the final code:

```
// Example 03C: Turn on LED when the button is pressed
// and keep it on after it is released
// including simple de-bouncing
// Now with another new and improved formula!!

#define LED  13   // the pin for the LED
#define BUTTON 7  // the input pin where the
                  // pushbutton is connected
int val = 0;      // val will be used to store the state
                  // of the input pin
int old_val = 0;  // this variable stores the previous
                  // value of "val"
int state = 0;    // 0 = LED off and 1 = LED on

void setup() {
  pinMode(LED, OUTPUT);    // tell Arduino LED is an output
  pinMode(BUTTON, INPUT);  // and BUTTON is an input
}

void loop(){
  val = digitalRead(BUTTON); // read input value and store it
                             // yum, fresh

  // check if there was a transition
  if ((val == HIGH) && (old_val == LOW)){
    state = 1 - state;
    delay(10);
  }

  old_val = val; // val is now old, let's store it

  if (state == 1) {
    digitalWrite(LED, HIGH); // turn LED ON
  } else {
    digitalWrite(LED, LOW);
  }
}
```

5/Advanced Input and Output

What you have just learned in Chapter 4 are the most elementary operations we can do in Arduino: controlling digital output and reading digital input. If Arduino were some sort of human language, those would be two letters of its alphabet. Considering that there are just five letters in this alphabet, you can see how much more work we have to do before we can write Arduino poetry.

Trying Out Other On/Off Sensors

Now that you've learned how to use a pushbutton, you should know that there are many other very basic sensors that work according to the same principle:

Switches
Just like a pushbutton, but doesn't automatically change state when released

Thermostats
A switch that opens when the temperature reaches a set value

Magnetic switches (also known as "reed relays")
Has two contacts that come together when they are near a magnet; used by burglar alarms to detect when a window is opened

Carpet switches
Small mats that you can place under a carpet or a doormat to detect the presence of a human being (or heavy cat)

Tilt switches

A simple electronic component that contains two contacts and a little metal ball (or a drop of mercury, but I don't recommend using those) An example of a tilt switch is called a tilt sensor. Figure 5-1 shows the inside of a typical model. When the sensor is in its upright position, the ball bridges the two contacts, and this works just as if you had pressed a pushbutton. When you tilt this sensor, the ball moves, and the contact is opened, which is just as if you had released a pushbutton. Using this simple component, you can implement, for example, gestural interfaces that react when an object is moved or shaken.

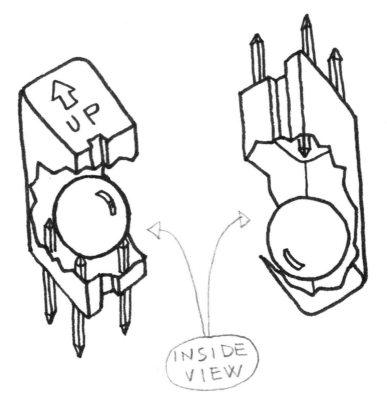

Figure 5-1.
The inside of a tilt sensor

Another sensor that you might want to try is the infrared sensor as found in burglar alarms (also known as a passive infrared or PIR sensor; see Figure 5-2). This small device triggers when a human being (or other living being) moves within its proximity. It's a simple way to detect motion.

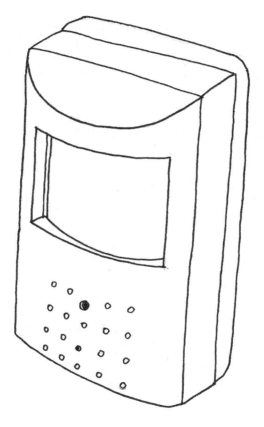

Figure 5-2.
Typical PIR sensor

You should now experiment by looking at all the possible devices that have two contacts that close, like the thermostat that sets a room's temperature (use an old one that's no longer connected), or just placing two contacts next to each other and dropping water onto them.

For example, by using the final example from Chapter 4 and a PIR sensor, you could make your lamp respond to the presence of human beings, or you could use a tilt switch to build one that turns off when it's tilted on one side.

Controlling Light with PWM

With the knowledge that you have so far gained, you could build an interactive lamp that can be controlled—and not just with a boring on/off switch, but maybe in a way that's a bit more elegant. One of the limitations of the blinking LED examples that we have used so far is that you can turn the light only on and off. A fancy interactive lamp needs to be dimmable. To solve this problem, we can use a little trick that makes a lot of things such as TV or cinema possible: persistence of vision.

As I hinted after the first example in Chapter 4, if you change the numbers in the delay function until you don't see the LED blinking any more, you will notice that the LED seems to be dimmed at 50% of its normal brightness. Now change the numbers so that the LED is on is one quarter of the time that it's off. Run the sketch and you'll see that the brightness is roughly 25%. This technique is called **pulse width modulation (PWM)**, a fancy way of saying that if you blink the LED fast enough, you don't see it blink any more, but you can change its brightness by changing the ratio between the on time and the off time. Figure 5-3 shows how this works.

This technique also works with devices other than an LED. For example, you can change the speed of a motor in the same way.

While experimenting, you will see that blinking the LED by putting delays in your code is a bit inconvenient, because as soon as you want to read a sensor or send data on the serial port, the LED will flicker while it's waiting for you to finish reading the sensor. Luckily, the processor used by the Arduino board has a piece of hardware that can very efficiently blink three LEDs while your sketch does something else. This hardware is implemented in pins 9, 10, and 11, which can be controlled by the *analogWrite()* instruction.

PWM

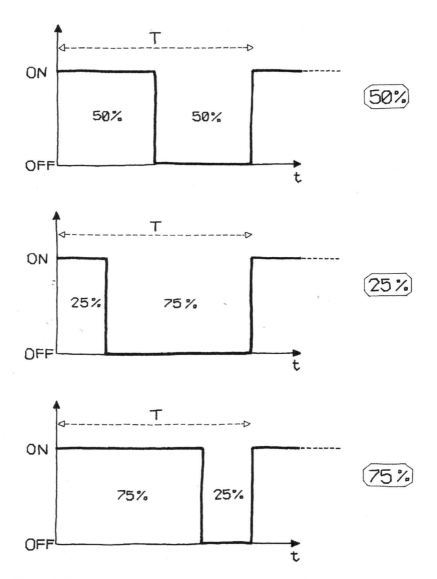

Figure 5-3.
PWM in action

For example, writing *analogWrite(9,128)* will set the brightness of an LED connected to pin 9 to 50%. Why 128? *analogWrite()* expects a number between 0 and 255 as an argument, where 255 means full brightness and 0 means off.

NOTE: Having three channels is very good, because if you buy red, green, and blue LEDs, you can mix their lights and make light of any colour that you like!

Let's try it out. Build the circuit that you see in Figure 5-4. Note that LEDs are polarized: the long pin (positive) should go to the right, and the short pin (negative) to the left. Also, most LEDs have a flattened negative side, as shown in the figure.

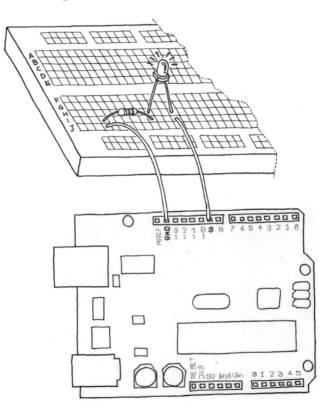

Figure 5-4:
LED connected to PWM pin

Then, create a new sketch in Arduino and use Example 04 (you can also download code examples from **www.makezine.com/getstartedarduino**):

```
// Example 04: Fade an LED in and out like on
// a sleeping Apple computer

#define LED    9 // the pin for the LED
int i = 0;      // We'll use this to count up and down

void setup() {
  pinMode(LED, OUTPUT); // tell Arduino LED is an output
}

void loop(){

  for (i = 0; i < 255; i++) { // loop from 0 to 254 (fade in)
    analogWrite(LED, i);       // set the LED brightness
    delay(10); // Wait 10ms because analogWrite
               // is instantaneous and we would
               // not see any change
  }

  for (i = 255; i > 0; i--) { // loop from 255 to 1 (fade out)

    analogWrite(LED, i); // set the LED brightness
    delay(10);           // Wait 10ms
  }

}
```

Now you have a replicated a fancy feature of a laptop computer (maybe it's a bit of a waste to use Arduino for something so simple). Let's the use this knowledge to improve our lamp.

Add the circuit we used to read a button (back in Chapter 4) to this breadboard. See if you can do this without looking at the next page, because I want you to start thinking about the fact that each elementary circuit I show here is a "building block" to make bigger and bigger projects. If you need to peek ahead, don't worry; the most important thing is that you spend some time thinking about how it might look.

To create this circuit, you will need to combine the circuit you just built (shown in Figure 5-4) with the pushbutton circuit shown in Figure 4-6. If you'd like, you can simply build both circuits on different parts of the breadboard; you have plenty of room. However, one of the advantages of the breadboard (see Appendix A) is that there is a pair of rails running horizontally across the bottom and top. One is marked red (for positive) and the other blue or black (for ground).

These rails are used to distribute power and ground to where it's needed. In the case of the circuit you need to build for this example, you have two components (both of them resistors) that need to be connected to the GND (ground) pin on the Arduino. Because the Arduino has two GND pins, you could simply connect these two circuits exactly as shown in each of the two figures; just hook them both up to the Arduino at the same time. Or, you could connect one wire from the breadboard's ground rail to one of the GND pins on the Arduino, and then take the wires that are connected to GND in the figures and connect them instead to the breadboard ground rail.

If you're not ready to try this, don't worry: simply wire up both circuits to your Arduino as shown in Figures 4-6 and 5-4. You'll see an example that uses the ground and positive breadboard rails in Chapter 6.

Getting back to this next example, if we have just one pushbutton, how do we control the brightness of a lamp? We're going to learn yet another interaction design technique: detecting how long a button has been pressed. To do this, I need to upgrade example 03C from Chapter 4 to add dimming. The idea is to build an "interface" in which a press and release action switches the light on and off, and a press and hold action changes brightness.

Let's have a look at the sketch:

```
//  Example 05: Turn on LED when the button is pressed
//  and keep it on after it is released
//  including simple de-bouncing.
//  If the button is held, brightness changes.

#define LED 9      // the pin for the LED
#define BUTTON 7  // input pin of the pushbutton

int val = 0;      // stores the state of the input pin

int old_val = 0; // stores the previous value of "val"
int state = 0;    // 0 = LED off while 1 = LED on

int brightness = 128;        // Stores the brightness value
unsigned long startTime = 0; // when did we begin pressing?

void setup() {
  pinMode(LED, OUTPUT);    // tell Arduino LED is an output
  pinMode(BUTTON, INPUT); // and BUTTON is an input
}

void loop() {

  val = digitalRead(BUTTON); // read input value and store it
                             // yum, fresh

  // check if there was a transition
  if ((val == HIGH) && (old_val == LOW)) {

    state = 1 - state; // change the state from off to on
                       // or vice-versa

    startTime = millis(); // millis() is the Arduino clock
                          // it returns how many milliseconds
                          // have passed since the board has
                          // been reset.

    // (this line remembers when the button
    // was last pressed)
    delay(10);
  }
```

```
// check whether the button is being held down
if ((val == HIGH) && (old_val == HIGH)) {

  // If the button is held for more than 500ms.
  if (state == 1 && (millis() - startTime) > 500) {

    brightness++; // increment brightness by 1
    delay(10);    // delay to avoid brightness going
                  // up too fast

    if (brightness > 255) { // 255 is the max brightness

      brightness = 0; // if we go over 255
                      // let's go back to 0
    }
  }
}

old_val = val; // val is now old, let's store it

if (state == 1) {
  analogWrite(LED, brightness); // turn LED ON at the
                                // current brightness level
} else {
  analogWrite(LED, 0); // turn LED OFF
}
}
```

Now try it out. As you can see, our interaction model is taking shape. If you press the button and release it immediately, you switch the lamp on or off. If you hold the button down, the brightness changes; just let go when you have reached the desired brightness.

Now let's learn how to use some more interesting sensors.

Use a Light Sensor Instead of the Pushbutton

Now we're going to try an interesting experiment. Take a light sensor, like the one pictured in Figure 5-5. You can get a five-pack of these from RadioShack (part number 276–1657).

Figure 5-5.
Light-dependent resistor (LDR)

In darkness, the resistance of a light-dependent resistor (LDR) is quite high. When you shine some light at it, the resistance quickly drops and it becomes a reasonably good conductor of electricity. It is thus a kind of light-activated switch.

Build the circuit that came with Example 02 (see "Using a Pushbutton to Control the LED" in Chapter 4), then upload the code from Example 02 to your Arduino.

Now plug the LDR onto the breadboard instead of the pushbutton. You will notice that if you cover the LDR with your hands, the LED turns off. Uncover the LDR, and the light goes on. You've just built your first real sensor-driven LED. This is important because for the first time in this book, we are using an electronic component that is not a simple mechanical device: it's a real rich sensor.

Analogue Input

As you learned in the previous section, Arduino is able to detect whether there is a voltage applied to one of its pins and report it through the *digitalRead()* function. This kind of either/or response is fine in a lot of applications, but the light sensor that we just used is able to tell us not just whether there is light, but also how much light there is. This is the difference between an on/off sensor (which tells us whether something is there) and an analogue sensor, whose value continuously changes. In order to read this type of sensor, we need a different type of pin.

In the lower-right part of the Arduino board, you'll see six pins marked "Analog In"; these are special pins that can tell us not only whether there is a voltage applied to them, but if so, also its value. By using the *analogRead()* function, we can read the voltage applied to one of the pins. This function returns a number between 0 and 1023, which represents voltages between 0 and 5 volts. For example, if there is a voltage of 2.5 V applied to pin number 0, *analogRead(0)* returns 512.

If you now build the circuit that you see in Figure 5-6, using a 10k resistor, and run the code listed in Example 06A, you'll see the onboard LED (you could also insert your own LED into pins 13 and GND as shown in "Blinking an LED" in Chapter 4) blinking at a rate that's dependent upon the amount of light that hits the sensor.

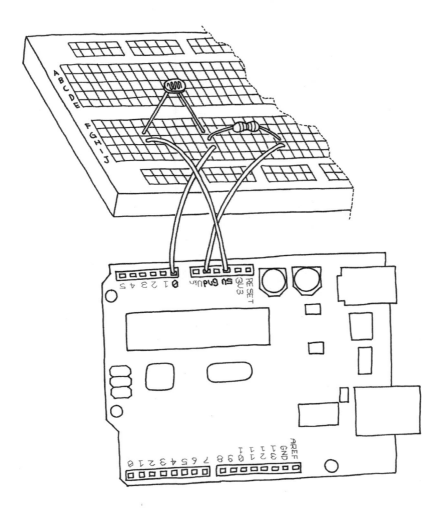

Figure 5-6.
An analogue sensor circuit

```
// Example 06A: Blink LED at a rate specified by the
// value of the analogue input

#define LED  13 // the pin for the LED

int val = 0;   // variable used to store the value
               // coming from the sensor
void setup() {
  pinMode(LED, OUTPUT); // LED is as an OUTPUT

  // Note: Analogue pins are
  // automatically set as inputs
}

void loop() {

  val = analogRead(0); // read the value from
                       // the sensor

  digitalWrite(13, HIGH); // turn the LED on

  delay(val); // stop the program for
              // some time

  digitalWrite(13, LOW); // turn the LED off

  delay(val); // stop the program for
              // some time
}
```

Now, try Example 06B: but before you do, you'll need to modify your circuit. Take a look at Figure 5-4 again and hook the LED up to pin 9 as shown. Because you've already got some stuff on the breadboard, you'll need to find a spot on the breadboard where the LED, wires, and resistor won't overlap with the LDR circuit.

```
// Example 06B: Set the brightness of LED to
// a brightness specified by the
// value of the analogue input

#define LED  9  // the pin for the LED

int val = 0;    // variable used to store the value
                // coming from the sensor

void setup() {

  pinMode(LED, OUTPUT); // LED is as an OUTPUT

  // Note: Analogue pins are
  // automatically set as inputs
}

void loop() {

  val = analogRead(0); // read the value from
                       // the sensor
  analogWrite(LED, val/4); // turn the LED on at
                           // the brightness set
                           // by the sensor

  delay(10); // stop the program for
             // some time
}
```

NOTE: we specify the brightness by dividing *val* by 4, because
analogRead() returns a number up to 1023, and *analogWrite()*
accepts a maximum of 255.

Try Other Analogue Sensors

Using the same circuit that you have seen in the previous section, you can connect a lot of other resistive sensors that work in more or less the same way. For example, you could connect a thermistor, which is a simple device whose resistance changes with temperature. In the circuit, I have shown you how changes in resistance become changes in voltage that can be measured by Arduino.

If you do work with a thermistor, be aware that there isn't a direct connection between the value you read and the actual temperature measured. If you need an exact reading, you should read the numbers that come out of the analogue pin while measuring with a real thermometer. You could put these numbers side by side in a table and work out a way to calibrate the analogue results to real-world temperatures.

Until now, we have just used an LED as an output device, but how do we read the actual values that Arduino is reading from the sensor? We can't make the board blink the values in Morse code (well, we could, but there is an easier way for humans to read the values). For this, we can have Arduino talk to a computer over a serial port, which is described in the next section.

Serial Communication

You learned at the beginning of this book that Arduino has a USB connection that is used by the IDE to upload code into the processor. The good news is that this connection can also be used by the sketches that we write in Arduino to send data back to the computer or to receive commands from it. For this purpose, we are going to use a serial object (an **object** is a collection of capabilities that are bundled together for the convenience of people writing sketches).

This object contains all the code that we need to send and receive data. We're now going to use the last circuit we built with the photoresistor and send the values that are read back to the computer. Type this code into a new sketch (you can also download the code from **www.makezine.com/ getstartedarduino**):

```
// Example 07: Send to the computer the values read from
// analogue input 0
// Make sure you click on "Serial Monitor"
// after you upload

#define SENSOR 0  // select the input pin for the
                  // sensor resistor

int val = 0; // variable to store the value coming
             // from the sensor

void setup() {

  Serial.begin(9600); // open the serial port to send
                      // data back to the computer at
                      // 9600 bits per second
}

void loop() {

  val = analogRead(SENSOR); // read the value from
                            // the sensor

  Serial.println(val); // print the value to
                       // the serial port

  delay(100); // wait 100ms between
              // each send
}
```

After you've uploaded the code to your Arduino, press the "Serial Monitor" button on the Arduino IDE (the rightmost button in the tool-bar); you'll see the numbers rolling past in the bottom part of the window. Now, any software that can read from the serial port can talk to Arduino. There are many programming languages that let you write programs on your computer that can talk to the serial port. Processing (www.processing.org) is a great complement to Arduino, because the languages and IDEs are so similar.

Driving Bigger Loads (Motors, Lamps, and the Like)

Each one of the pins on an Arduino board can be used to power devices that use up to 20 milliamps: this is a very small amount of current, just enough to drive an LED. If you try to drive something like a motor, the pin will immediately stop working, and could potentially burn out the whole processor. To drive bigger loads like motors or incandescent lamps, we need to use an external component that can switch such things on and off and that is driven by an Arduino pin. One such device is called a MOSFET transistor—ignore the funny name—it's an electronic switch that can be driven by applying a voltage to one of its three pins, each of which is called a *gate*. It is something like the light switch that we use at home, where the action of a finger turning the light on and off is replaced by a pin on the Arduino board sending voltage to the gate of the MOSFET.

NOTE: MOSFET means "metal–oxide–semiconductor field-effect transistor." It's a special type of transistor that operates based on the field-effect principle. This means that electricity will flow though a piece of semiconductor material (between the Drain and Source pins) when a voltage is applied to the Gate pin. As the Gate is insulated from the rest through a layer of metal oxide, there is no current flowing from Arduino into the MOSFET, making it very simple to interface. They are ideal for switching on and off large loads at high frequencies.

In Figure 5-7, you can see how you would use a MOSFET like the IRF520 to turn on and off a small motor attached to a fan. You will also notice that the motor takes its power supply from the 9 V connector on the Arduino board. This is another benefit of the MOSFET: it allows us to drive devices whose power supply differs from the one used by Arduino. As the MOSFET is connected to pin 9, we can also use *analogWrite()* to control the speed of the motor through PWM.

Complex Sensors

We define **complex sensors** as those that produce a type of information that requires a bit more than a *digitalRead()* or an *analogRead()* function to be used. These usually are small circuits with a small microcontroller inside that preprocesses the information.

Some of the complex sensors available include ultrasonic rangers, infrared rangers, and accelerometer. You can find examples on how to use them on our website in the "Tutorials" section (**www.arduino.cc/en/**Tutorial/HomePage).

Tom Igoe's *Making Things Talk* (O'Reilly) has extensive coverage of these sensors and many other complex sensors.

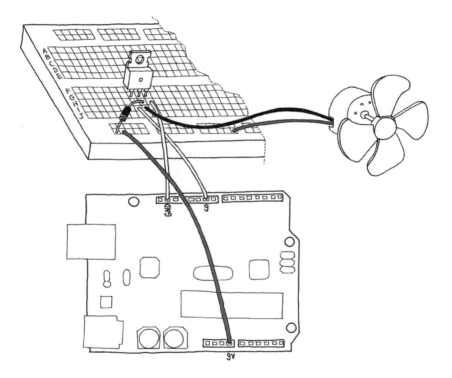

Figure 5-7.
A motor circuit for Arduino

6/Talking to the Cloud

In the preceding chapters, you learned the basics of Arduino and the fundamental building blocks available to you. Let me remind you what makes up the "Arduino Alphabet":

Digital Output
We used it to control an LED but, with the proper circuit, it can be used to control motors, make sounds, and a lot more.

Analog Output
This gives us the ability to control the brightness of the LED, not just turn it on or off. We can even control the speed of a motor with it.

Digital Input
This allows us to read the state of simple sensors, like pushbuttons or tilt switches.

Analog Input
We can read signals from sensors that send a continuous signal that's not just on or off, such as a potentiometer or a light sensor.

Serial Communication
This allows us to communicate with a computer and exchange data or simply monitor what's going on with the sketch that's running on the Arduino.

In this chapter, we're going to see how to put together a working application using what you have learned in the previous chapters. This chapter should show you how every single example can be used as a building block for a complex project.

Here is where the wannabe designer in me comes out. We're going to make the twenty-first-century version of a classic lamp by my favourite Italian designer, Joe Colombo. The object we're going to build is inspired by a lamp called "Aton" from 1964.

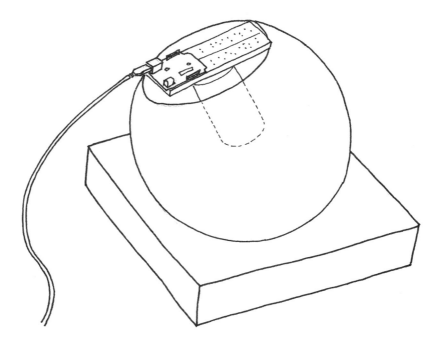

Figure 6-1.
The finished lamp

The lamp, as you can see in Figure 6-1, is a simple sphere sitting on a base with a large hole to keep the sphere from rolling off your desk. This design allows you to orient the lamp in different directions.

In terms of functionality, we want to build a device that would connect to the Internet, fetch the current list of articles on the Make blog (**blog. makezine.com**) and count how many times the words "peace", "love", and "Arduino" are mentioned. With these values, we're going to generate a colour and display it on the lamp. The lamp itself has a button we can use to turn it on and off, and a light sensor for automatic activation.

Planning

Let's look at what we want to achieve and what bits and pieces we need. First of all, we need Arduino to be able to connect to the Internet. As the Arduino board has only a USB port, we can't plug it directly into an Internet connection, so we need to figure out how to bridge the two. Usually what people do is run an application on a computer that will connect to the Internet, process the data, and send Arduino some simple bit of distilled information.

Arduino is a simple computer with a small memory; it can't process large files easily, and when we connect to an RSS feed we'll get a very verbose XML file that would require a lot more RAM. We'll implement a *proxy* to simplify the XML using the Processing language.

Processing

Processing is where Arduino came from. We love this language and use it to teach programming to beginners as well as to build beautiful code. Processing and Arduino are the perfect combination. Another advantage is that Processing is open source and runs on all the major platforms (Mac, Linux, and Windows). It can also generate stand-alone applications that run on those platforms. What's more, the Processing community is lively and helpful, and you can find thousands of premade example programs.

The proxy does the following work for us: it downloads the RSS feed from **makezine.com** and extracts all the words from the resulting XML file. Then, going through all of them, it counts the number of times "peace", "love", and "Arduino" appear in the text. With these three numbers, we'll calculate a colour value and send it to Arduino. The board will send back the amount of light measured by the sensor and show it on the computer screen.

On the hardware side, we'll combine the pushbutton example, the light sensor example, the PWM LED control (multiplied by 3!) and serial communication.

As Arduino is a simple device, we'll need to codify the colour in a simple way. We'll use the standard way that colours are represented in HTML: # followed by six hexadecimal digits.

Hexadecimal numbers are handy, because each 8-bit number is stored in exactly two characters; with decimal numbers this varies from one to three characters. Predictability also makes the code simpler: we wait until we see an #, then we read the six characters that follow into a *buffer* (a variable used as a temporary holding area for data). Finally, we turn each group of two characters into a byte that represents the brightness of one of the three LEDs.

Coding

There are two sketches that you'll be running: one Processing sketch, and one Arduino sketch. Here is the code for the Processing sketch. You can download it from **www.makezine.com/getstartedarduino**.

```
// Example 08A: Arduino networked lamp
// parts of the code are inspired
// by a blog post by Tod E. Kurt (todbot.com)

import processing.serial.*;

String feed = "http://blog.makezine.com/index.xml";

int interval = 10;  // retrieve feed every 60 seconds;
int lastTime;       // the last time we fetched the content

int love  = 0;
int peace = 0;
int arduino = 0;

int light = 0;  // light level measured by the lamp

Serial port;
color c;
String cs;

String buffer = ""; // Accumulates characters coming from Arduino

PFont font;

void setup() {
  size(640,480);
  frameRate(10);   // we don't need fast updates

  font = loadFont("HelveticaNeue-Bold-32.vlw");
  fill(255);
  textFont(font, 32);
```

```
  // IMPORTANT NOTE:
  // The first serial port retrieved by Serial.list()
  // should be your Arduino. If not, uncomment the next
  // line by deleting the // before it, and re-run the
  // sketch to see a list of serial ports. Then, change
  // the 0 in between [ and ] to the number of the port
  // that your Arduino is connected to.
  //println(Serial.list());
  String arduinoPort = Serial.list()[0];
  port = new Serial(this, arduinoPort, 9600); // connect to Arduino

  lastTime = 0;
  fetchData();
}

void draw() {
  background( c );
  int n = (interval - ((millis()-lastTime)/1000));

  // Build a colour based on the 3 values
  c = color(peace, love, arduino);
  cs = "#" + hex(c,6); // Prepare a string to be sent to Arduino

  text("Arduino Networked Lamp", 10,40);
  text("Reading feed:", 10, 100);
  text(feed, 10, 140);

  text("Next update in "+ n + " seconds",10,450);
  text("peace" ,10,200);
  text(" " + peace, 130, 200);
  rect(200,172, peace, 28);

  text("love ",10,240);
  text(" " + love, 130, 240);
  rect(200,212, love, 28);

  text("arduino ",10,280);
  text(" " + arduino, 130, 280);
  rect(200,252, arduino, 28);

  // write the colour string to the screen
  text("sending", 10, 340);
  text(cs, 200,340);
```

```
  text("light level", 10, 380);
  rect(200, 352,light/10.23,28); // this turns 1023 into 100

  if (n <= 0) {
    fetchData();
    lastTime = millis();
  }


  port.write(cs); // send data to Arduino

  if (port.available() > 0) { // check if there is data waiting
    int inByte = port.read(); // read one byte
    if (inByte != 10) { // if byte is not newline
      buffer = buffer + char(inByte); // just add it to the buffer
    }
    else {

      // newline reached, let's process the data
      if (buffer.length() > 1) { // make sure there is enough data

        // chop off the last character, it's a carriage return
        // (a carriage return is the character at the end of a
        // line of text)
        buffer = buffer.substring(0,buffer.length() -1);

        // turn the buffer from string into an integer number
        light = int(buffer);

        // clean the buffer for the next read cycle
        buffer = "";

        // We're likely falling behind in taking readings
        // from Arduino. So let's clear the backlog of
        // incoming sensor readings so the next reading is
        // up-to-date.
        port.clear();
      }
    }
  }

}
```

```
void fetchData() {
  // we use these strings to parse the feed
  String data;
  String chunk;

  // zero the counters
  love    = 0;
  peace   = 0;
  arduino = 0;
  try {
    URL url = new URL(feed);  // An object to represent the URL
    // prepare a connection
    URLConnection conn = url.openConnection();
    conn.connect(); // now connect to the Website

    // this is a bit of virtual plumbing as we connect
    // the data coming from the connection to a buffered
    // reader that reads the data one line at a time.
    BufferedReader in = new
      BufferedReader(new InputStreamReader(conn.getInputStream()));

    // read each line from the feed
    while ((data = in.readLine()) != null) {

      StringTokenizer st =
        new StringTokenizer(data,"\"<>,.()[] ");// break it down
      while (st.hasMoreTokens()) {
        // each chunk of data is made lowercase
        chunk= st.nextToken().toLowerCase() ;

        if (chunk.indexOf("love") >= 0 ) // found "love"?
          love++;    // increment love by 1
        if (chunk.indexOf("peace") >= 0)   // found "peace"?
          peace++;   // increment peace by 1
        if (chunk.indexOf("arduino") >= 0) // found "arduino"?
          arduino++; // increment arduino by 1
      }
    }

    // Set 64 to be the maximum number of references we care about.
    if (peace > 64)    peace = 64;
    if (love > 64)     love = 64;
    if (arduino > 64) arduino = 64;
```

```
    peace = peace * 4;       // multiply by 4 so that the max is 255,
    love = love * 4;         // which comes in handy when building a
    arduino = arduino * 4; // colour that is made of 4 bytes (ARGB)
  }
  catch (Exception ex) { // If there was an error, stop the sketch
    ex.printStackTrace();
    System.out.println("ERROR: "+ex.getMessage());
  }

}
```

There are two things you need to do before the Processing sketch will run correctly. First, you need to tell Processing to generate the font that we are using for the sketch. To do this, create and save this sketch. Then, with the sketch still opened, click Processing's Tools menu, then select Create Font. Select the font named HelveticaNeue-Bold, choose 32 for the font size, and then click OK.

Second, you will need to confirm that the sketch is using the correct serial port for talking to Arduino. You'll need to wait until you've assembled the Arduino circuit and uploaded the Arduino sketch before you can confirm this. On most systems, this Processing sketch will run fine. However, if you don't see anything happening on the Arduino and you don't see any information from the light sensor appearing onscreen, find the comment labeled "IMPORTANT NOTE" in the Processing sketch and follow the instructions there.

Here is the Arduino sketch (also available at **www.makezine.com/ getstartedarduino**):

```
// Example 08B: Arduino Networked Lamp
#define SENSOR 0
#define R_LED 9
#define G_LED 10
#define B_LED 11
#define BUTTON 12

int val = 0; // variable to store the value coming from the sensor

int btn = LOW;
int old_btn = LOW;
int state = 0;
char buffer[7] ;
int pointer = 0;
byte inByte = 0;

byte r = 0;
byte g = 0;
byte b = 0;

void setup() {
  Serial.begin(9600);  // open the serial port
  pinMode(BUTTON, INPUT);
}

void loop() {
  val = analogRead(SENSOR); // read the value from the sensor
  Serial.println(val);       // print the value to
                             // the serial port

  if (Serial.available() >0) {

    // read the incoming byte:
    inByte = Serial.read();

    // If the marker's found, next 6 characters are the colour
    if (inByte == '#') {

      while (pointer < 6) { // accumulate 6 chars
        buffer[pointer] = Serial.read(); // store in the buffer
        pointer++; // move the pointer forward by 1
      }
```

```
    // now we have the 3 numbers stored as hex numbers
    // we need to decode them into 3 bytes r, g and b
    r = hex2dec(buffer[1]) + hex2dec(buffer[0]) * 16;
    g = hex2dec(buffer[3]) + hex2dec(buffer[2]) * 16;
    b = hex2dec(buffer[5]) + hex2dec(buffer[4]) * 16;

    pointer = 0; // reset the pointer so we can reuse the buffer

  }
}

btn = digitalRead(BUTTON); // read input value and store it

// Check if there was a transition
if ((btn == HIGH) && (old_btn == LOW)){
  state = 1 - state;
}

old_btn = btn; // val is now old, let's store it

if (state == 1) { // if the lamp is on

  analogWrite(R_LED, r);  // turn the leds on
  analogWrite(G_LED, g);  // at the colour
  analogWrite(B_LED, b);  // sent by the computer
} else {

  analogWrite(R_LED, 0);  // otherwise turn off
  analogWrite(G_LED, 0);
  analogWrite(B_LED, 0);
 }

  delay(100);                    // wait 100ms between each send
}

int hex2dec(byte c) { // converts one HEX character into a number
    if (c >= '0' && c <= '9') {
      return c - '0';
    } else if (c >= 'A' && c <= 'F') {
      return c - 'A' + 10;
    }
}
```

Assembling the Circuit

Figure 6-2 shows how to assemble the circuit. You need to use 10K resistors for all of the resistors shown in the diagram, although you could get away with lower values for the resistors connected to the LEDs.

Remember from the PWM example in Chapter 5 that LEDs are polarized: in this circuit, the long pin (positive) should go to the right, and the short pin (negative) to the left. (Most LEDs have a flattened negative side, as shown in the figure.)

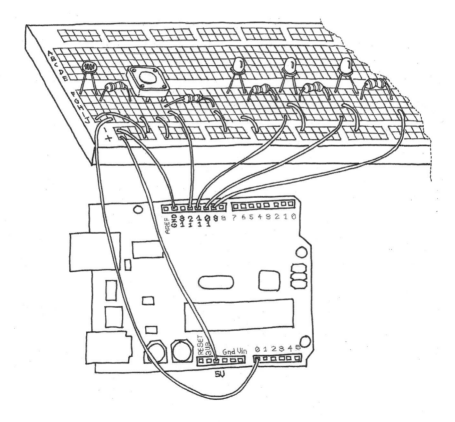

Figure 6-2.
The "Arduino Networked Lamp" circuit

Build the circuit as shown, using one red, one green, and one blue LED. Next, load the sketches into Arduino and Processing, then run the sketches and try it out. If you run into any problems, check Chapter 7, "Trouble-shooting".

Now let's complete the construction by placing the breadboard into a glass sphere. The simplest and cheapest way to do this is to buy an IKEA "FADO" table lamp. It's now selling for about US\$14.99/€14.99/£8.99 (ahh, the luxury of being European).

Instead of using three separate LEDs, you can use a single RGB LED, which has four leads coming off it. You'll hook it up in much the same way as the LEDs shown in Figure 6-2, with one change: instead of three separate connections to the ground pin on Arduino, you'll have a single lead (called the "common cathode") going to ground.

SparkFun sells a 4-lead RGB LED for a few dollars (**www.sparkfun.com**; part number COM-00105). Also, unlike discrete single-color LEDs, the longest lead on this RGB LED is the one that goes to ground. The three shorter leads will need to connect to Arduino pins 9, 10, and 11 (with a resistor between the leads and the pins, just as with the separate red, green, and blue LEDs).

Here's How to Assemble It:

Unpack the lamp and remove the cable that goes into the lamp from the bottom. You will no longer be plugging this into the wall.

Strap the Arduino on a breadboard and hot-glue the breadboard onto the back of the lamp.

Solder longer wires to the RGB LED and glue it where the lightbulb used to be. Connect the wires coming from the LED to the breadboard (where it was connected before you removed it). Remember that you will only need one connection to ground if you are using a 4-lead RGB LED.

Either find a nice piece of wood with a hole that can be used as a stand for the sphere or just cut the top of the cardboard box that came with the lamp at approximately 5cm (or 2") and make a hole with a diameter that cradles the lamp. Reinforce the inside of the cardboard box by using hot glue all along the inside edges, which will make the base more stable.

Place the sphere on the stand and bring the USB cable out of the top and connect it to the computer.

Fire off your Processing code, press the on/off button, and watch the lamp come to life.

As an exercise, try to add code that will turn on the lamp when the room gets dark. Other possible enhancements are:

» Add tilt sensors to turn the lamp on or off by rotating it in different directions.

» Add a small PIR sensor to detect when somebody is around and turn it off when nobody is there to watch.

» Create different modes so that you can get manual control of the colour or make it fade through many colours.

Think of different things, experiment, and have fun!

7/Troubleshooting

There will come a moment in your experimentation when nothing will be working and you will have to figure out how to fix it. Troubleshooting and debugging are ancient arts in which there are a few simple rules, but most of the results are obtained through a lot of work.

The more you work with electronics and Arduino, the more you will learn and gain experience, which will ultimately make the process less painful. Don't be discouraged by the problems that you will find—it's all easier than it seems at the beginning.

As every Arduino-based project is made both of hardware and software, there will be more than one place to look if something goes wrong. While looking for a bug, you should operate along three lines:

Understanding
Try to understand as much as possible how the parts that you're using work and how they're supposed to contribute to the finished project. This approach will allow you to devise some way to test each component separately.

Simplification and segmentation
The Ancient Romans used to say *divide et impera*: divide and rule. Try to break down (mentally) the project into its components by using the understanding you have and figure out where the responsibility of each component begins and ends.

Exclusion and certainty
While investigating, test each component separately so that you can be absolutely certain that each one works by itself. You will gradually build up confidence about which parts of project are doing their job and which ones are dubious.

Debugging is the term used to describe this process as applied to software. The legend says it was used for the first time by Grace Hopper back in the 1940s, when computers where mostly electromechanical, and one of them stopped working because actual insects got caught in the mechanisms.

Many of today's bugs are not physical anymore: they're virtual and invisible, at least in part. Therefore they require a sometimes lengthy and boring process to be identified.

Testing the Board

What if the very first example, "blink an LED," didn't work? Wouldn't that be a bit depressing? Let's figure out what to do.

Before you start blaming your project, you should make sure that a few things are in order, as airline pilots do when they go through a checklist to make sure that the airplane will be flying properly before takeoff:

Plug your Arduino into a USB plug on your computer.

» Make sure the computer is on (yes, it sounds silly, but it has happened). If the green light marked PWR turns on, this means that the computer is powering the board. If the LED seems very faint, something is wrong with the power: try a different USB cable and inspect the computer's USB port and the Arduino's USB plug to see whether there is any damage. If all else fails, try a different USB port on your computer or a different computer entirely.

» If the Arduino is brand new, the yellow LED marked L will start blinking in a bit of a nervous pattern; this is the test program that was loaded at the factory to test the board.

» If you have been using an external power supply and are using an old Arduino (Extreme, NG, or Diecimila), make sure that the power supply is plugged in and that the jumper marked SV1 is connecting the two pins that are nearest to the external power supply connector.

NOTE: When you are having trouble with other sketches and need to confirm that the board is functioning, open the first "blink an LED" example in the Arduino IDE and upload it to the board. The on-board LED should blink in a regular pattern.

If you have gone through all these steps successfully, then you can be confident that your Arduino is working correctly.

Testing Your Breadboarded Circuit

Now connect the board to your breadboard by running a jumper from the 5 V and GND connections to the positive and negative rails of the bread-board. If the green PWR LED turns off, remove the wires immediately. This means there is a big mistake in your circuit and you have a "short circuit" somewhere. When this happens, your board draws too much current and the power gets cut off to protect the computer.

NOTE: If you're a concerned that you may damage your computer, remember that on many computers, the current protection is usually quite good and responds quickly. Also, the Arduino board is fitted with a "PolyFuse," a current-protection device that resets itself when the fault is removed.

If you're really paranoid, you can always connect the Arduino board through a self-powered USB hub. In this case, if it all goes horribly wrong, the USB hub is the one that will be pushing up daisies, not your computer.

If you're getting a short circuit, you have to start the "simplification and segmentation" process. What you must do is go through every sensor in the project and connect just one at a time.

The first thing to start from is always the power supply (the connections from 5 V and GND). Look around and make sure that each part of the circuit is powered properly.

Working step by step and making one single modification at a time is the number one rule for fixing stuff. This rule was hammered into my young head by my school professor and first employer, Maurizio Pirola. Every time I'm debugging something and things don't look good (and believe me, it happens a lot), his face pops in my head saying "one modification at a time . . . one modification at a time" and that's usually when I fix everything. This is very important, because you will know what fixed the problem (it's all too easy to lose track of which modification actually solved the problem, which is why it's so important to make one at a time).

Each debugging experience will build up in your head a "knowledge base" of defects and possible fixes. And before you know it, you'll become an expert. This will make you look very cool, because as soon as a newbie says "This doesn't work!" you'll give it a quick look and have the answer in a split second.

Isolating Problems

Another important rule is to find a reliable way to reproduce a problem. If your circuit behaves in a funny way at random times, try really hard to figure out the exact moment the problem occurs and what is causing it. This process will allow you to think about a possible cause. It is also very useful when you need to explain to somebody else what's going on.

Describing the problem as precisely as possible is also a good way to find a solution. Try to find somebody to explain the problem to—in many cases, a solution will pop into your head as you articulate the problem. Brian W. Kernighan and Rob Pike, in *The Practice of Programming* (Addison-Wesley, 1999), tell the story of one university that "kept a teddy bear near the help desk. Students with mysterious bugs were required to explain them to the bear before they could speak to a human counselor."

Problems with the IDE

In some cases, you may have a problem using the Arduino IDE, particularly on Windows.

If you get an error when you double-click on the Arduino icon, or if nothing happens, try double-clicking the *run.bat* file as an alternative method to launch Arduino.

Windows users may also run into a problem if the operating system assigns a COM port number of COM10 or greater to Arduino. If this happens, you can usually convince Windows to assign a lower port number to Arduino. First, open up the Device Manager by clicking the Start menu, right-clicking on Computer (Vista) or My Computer (XP), and choosing Properties. On Windows XP, click Hardware and choose Device Manager. On Vista, click on Device Manager (it appears in the list of tasks on the left of the window).

Look for the serial devices in the list under "Ports (COM & LPT)." Find a serial device that you're not using that is numbered COM9 or lower. Right-click it and choose Properties from the menu. Then, choose the Port Settings tab and click Advanced. Set the COM port number to COM10 or higher, click OK and click OK again to dismiss the Properties dialog.

Now, do the same with the USB Serial Port device that represents Arduino, with one change: assign it the COM port number (COM9 or lower) that you just freed up.

If these suggestions don't help, or if you're having a problem not described here, check out the Arduino troubleshooting page at www.arduino.cc/en/Guide/Troubleshooting.

How to Get Help Online

If you are stuck, don't spend days running around alone—ask for help. One of the best things about Arduino is its community. You can always find help if you can describe your problem well.

Get the habit of cutting and pasting things into a search engine and see whether somebody is talking about it. For example, when the Arduino IDE spits out a nasty error message, copy and paste it into a Google search and see what comes out. Do the same with bits of code you're working on or just a specific function name. Look around you: everything has been invented already and it's stored somewhere on a web page.

For further investigation, start from the www.arduino.cc main website and look at the FAQ (www.arduino.cc/en/Main/FAQ), then move on to the playground (www.arduino.cc/playground), a freely editable wiki that any user can modify to contribute documentation. It's one of the best parts of the whole open source philosophy. People contribute documentation and examples of anything you can do with Arduino. Before you start a project, search the playground and you'll find a bit of code or a circuit diagram to get you started.

If you still can't find an answer that way, search the forum (www.arduino.cc/cgi-bin/yabb2/YaBB.pl). If that doesn't help, post a question there. Pick the correct board for your problem: there are different areas for software or hardware issues and even forums in five different languages. Please post as much information as you can:

» What Arduino board are you using?

» What operating system are you using to run the Arduino IDE?

» Give a general description of what you're trying to do. Post links to datasheets of strange parts you're using.

The number of answers you get depends on how well you formulate your question.

Your chances increase if you **avoid these things at all cost** (these rules are good for any online forums, not just Arduino's):

» Typing your message all in CAPITALS. It annoys people a lot and is like walking around with "newbie" tattooed on your forehead (in on-line communities, typing in all capitals is considered "shouting").

» Posting the same message in several different parts of the forum.

» "Bumping" your message by posting follow-up comments asking "Hey, how come no one replied?" or even worse, simply posting the text "bump." If you didn't get a reply, take a look at your posting. Was the subject clear? Did you provide a well-worded description of the problem you are having? Were you nice? Always be nice.

» Writing messages like "I want to build a space shuttle using arduino how do I do that". This means that you want people to do your work for you, and this approach is simply not fun for a real tinkerer. It's better to explain what you want to build and then ask a specific question about one part of the project and take it from there.

» A variation of the previous point is when the question is clearly some-thing the poster of the message is getting paid to do. If you ask specific questions people are happy to help, but if you ask them to do all your work (and you don't share the money), the response is likely to be less nice.

» Posting messages that look suspiciously like school assignments and asking the forum to do your homework. Professors like me roam the forums and slap such students with a large trout.

Appendix A/
The Breadboard

The process of getting a circuit to work involves making lots of changes to it until it behaves properly; it's a very fast, iterative process that's something like an electronic equivalent to sketching. The design evolves in your hands as you try different combinations. For the best results, use a system that allows you to change the connections between components in the fastest, most practical, and least destructive way. These requirements clearly rule out soldering, which is a time-consuming procedure that puts components under stress every time you heat them up and cool them down.

The answer to this problem is a very practical device called the solderless breadboard. As you can see from Figure A-1, it's a small plastic board full of holes, each of which contains a spring-loaded contact. You can push a component's leg into one of the holes, and it will establish an electrical connection with all of the other holes in the same vertical column of holes. Each hole is a distance of 2.54 mm from the others.

Because most of the components have their legs (known to techies as "pins") spaced at that standard distance, chips with multiple legs fit nicely. Not all of the contacts on a breadboard are created equal—there are some differences. The top and bottom rows (coloured in red and blue and marked with + and –) are connected horizontally and are used to carry the power across the board so that when you need power or ground, you can provide it very quickly with a *jumper* (a short piece of wire used to connect two points in the circuits). The last thing you need to know about breadboards is that in the middle, there is a large gap that is as wide as the size of a small chip. Each vertical line of holes is interrupted in the middle, so that when you plug in a chip, you don't short-circuit the pins that are on the two sides of the chip. Clever, eh?

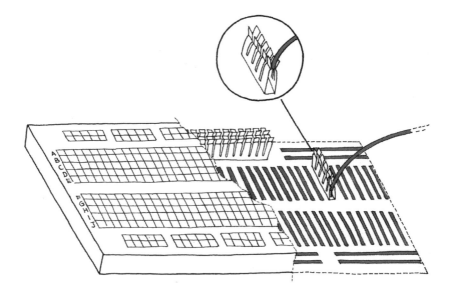

Figure A-1.
The solderless breadboard

Appendix B/Reading Resistors and Capacitors

In order to use electronic parts, you need to be able to identify them, which can be a difficult task for a beginner. Most of the resistors that you find in a shop have a cylindrical body with two legs sticking out and have strange coloured markings all around them. When the first commercial resistors were made, there was no way to print numbers small enough to fit on their body, so clever engineers decided that they could just represent the values with strips of coloured paint.

Today's beginners have to figure out a way to interpret these signs. The "key" is quite simple: generally, there are four stripes, and each colour represents a number. One of rings is usually gold-coloured; this one represents the precision of that resistor. To read the stripes in order, hold the resistor so the gold (or silver in some cases) stripe is to the right. Then, read the colours and map them to the corresponding numbers. In the following table, you'll find a translation between the colours and their numeric values.

Colour	Value
Black	0
Brown	1
Red	2
Orange	3
Yellow	4
Green	5
Blue	6
Purple	7
Grey	8
White	9
Silver	10%
Gold	5%

For example, brown, black, orange, and gold markings mean 1 0 3 ±5%. Easy, right? Not quite, because there is a twist: the third ring actually represents the number of zeros in the value. Therefore 1 0 3 is actually 1 0 followed by 3 zeros, so the end result is 10,000 ohms ±5%. Electronics geeks tend to shorten values by expressing them in kilo ohm (for thousands of ohms) and mega ohms (for millions of ohms), so a 10,000 ohm resistor is usually shortened to 10k, while 10,000,000 becomes 10M. Please note that because engineers are fond of optimising everything, on some schematic diagrams you might find values expressed as 4k7, which means 4.7 kilo ohms, or 4700.

Capacitors are a bit easier: the barrel-shaped capacitors (electrolytic capacitors) generally have their values printed on them. A capacitor's value is measured in farads (F), but most capacitors that you encounter will be measured in micro farads (µF). So if you see a capacitor labelled 100 µF, it's a 100 micro farad capacitor.

Many of the disc-shaped capacitors (ceramic capacitors) do not have their units listed, and use a three-digit numeric code indicating the number of pico farads (pF). There are 1,000,000 pF in one µF. Similar to the resistor codes, you use the third number to determine the number of zeros to put after the first two, with one difference: if you see 0–5, that indicates the number of zeros. 6 and 7 are not used, and 8 and 9 are handled differently. If you see 8, multiply the number that the first two digits form by 0.01, and if you see 9, multiply it by 0.1.

So, a capacitor labelled 104 would be 100,000 pF or 0.1 µF. A capacitor labeled 229 would be 2.2 pF.

Appendix C/Arduino Quick Reference

Here is a quick explanation of all the standard instructions supported by the Arduino language.

For a more detailed reference, see: arduino.cc/en/Reference/HomePage

STRUCTURE

An Arduino sketch runs in two parts:

```
void setup()
```
This is where you place the initialisation code—the instructions that set up the board before the main loop of the sketch starts.

```
void loop()
```
This contains the main code of your sketch. It contains a set of instructions that get repeated over and over until the board is switched off.

SPECIAL SYMBOLS

Arduino includes a number of symbols to delineate lines of code, comments, and blocks of code.

; (semicolon)

Every instruction (line of code) is terminated by a semicolon. This syntax lets you format the code freely. You could even put two instructions on the same line, as long as you separate them with a semicolon. (However, this would make the code harder to read.)

Example:
```
delay(100);
```

{} (curly braces)

This is used to mark blocks of code. For example, when you write code for the *loop()* function, you have to use curly braces before and after the code.

Example:
```
void loop() {
   Serial.println("ciao");
}
```

comments

These are portions of text ignored by the Arduino processor, but are extremely useful to remind yourself (or others) of what a piece of code does.

There are two styles of comments in Arduino:

```
// single-line: this text is ignored until the end of the line
/* multiple-line:
   you can write
   a whole poem in here
*/
```

CONSTANTS

Arduino includes a set of predefined keywords with special values. **HIGH** and **LOW** are used, for example, when you want to turn on or off an Arduino pin. **INPUT** and **OUTPUT** are used to set a specific pin to be either and input or an output

true and **false** indicate exactly what their names suggest: the truth or falsehood of a condition or expression.

VARIABLES

Variables are named areas of the Arduino's memory where you can store data that you can use and manipulate in your sketch. As the name suggests, they can be changed as many times as you like.

Because Arduino is a very simple processor, when you declare a variable you have to specify its type. This means telling the processor the size of the value you want to store.

Here are the *datatypes* that are available:

boolean

Can have one of two values: true or false.

char

Holds a single character, such as A. Like any computer, Arduino stores it as a number, even though you see text. When chars are used to store numbers, they can hold values from −128 to 127.

byte

Holds a number between 0 and 255. As with chars, bytes use only one byte of memory.

int

Uses 2 bytes of memory to represent a number between −32,768 and 32,767; it's the most common data type used in Arduino.

unsigned int

Like int, uses 2 bytes but the **unsigned** prefix means that it can't store negative numbers, so its range goes from 0 to 65,535.

long

This is twice the size of an **int** and holds numbers from −2,147,483,648 to 2,147,483,647.

unsigned long

Unsigned version of **long**; it goes from 0 to 4,294,967,295.

float

This quite big and can hold floating-point values, a fancy way of saying that you can use it to store numbers with a decimal point in it. It will eat up 4 bytes of your precious RAM and the functions that can handle them use up a lot of code memory as well. So use **float**s sparingly.

double

Double-precision floating-point number, with a maximum value of $1.7976931348623157 \times 10^{308}$. Wow, that's huge!

string

A set of ASCII characters that are used to store textual information (you might use a string to send a message via a serial port, or to display on an LCD display). For storage, they use one byte for each character in the string, plus a null character to tell Arduino that it's the end of the string. The following are equivalent:

```
char string1[]  = "Arduino"; // 7 chars + 1 null char
char string2[8] = "Arduino"; // Same as above
```

array
A list of variables that can be accessed via an index. They are used to build tables of values that can easily be accessed. For example, if you want to store different levels of brightness to be used when fading an LED, you could create six variables called light01, light02, and so on. Better yet, you could use a simple array like:

```
int light[6] = {0, 20, 50, 75, 100};
```

The word "array" is not actually used in the variable declaration: the symbols [] and {} do the job.

CONTROL STRUCTURES
Arduino includes keywords for controlling the logical flow of your sketch.

if ... else
This structure makes decisions in your program. *if* must be followed by a question specified as an expression contained in parentheses. If the expression is true, whatever follows will be executed. If it's false, the block of code following else will be executed. It's possible to use just *if* without providing an *else* clause.

Example:
```
if (val == 1) {
  digitalWrite(LED,HIGH);
}
```

for
Lets you repeat a block of code a specified number of times.

Example:
```
for (int i = 0; i < 10; i++) {
    Serial.print("ciao");
}
```

switch case
The *if* statement is like a fork in the road for your program. *switch case* is like a massive roundabout. It lets your program take a variety of directions

depending on the value of a variable. It's quite useful to keep your code tidy as it replaces long lists of *if* statements.

Example:
```
switch (sensorValue) {
    case 23:
        digitalWrite(13,HIGH);
        break;
    case 46:
        digitalWrite(12,HIGH);
        break;
    default: // if nothing matches this is executed
        digitalWrite(12,LOW);
        digitalWrite(13,LOW);
}
```

while
Similar to *if*, this executes a block of code while a certain condition is true.

Example:
```
// blink LED while sensor is below 512
sensorValue = analogRead(1);
while (sensorValue < 512) {
    digitalWrite(13,HIGH);
    delay(100);
    digitalWrite(13,HIGH);
    delay(100);
    sensorValue = analogRead(1);
}
```

do . . . while
Just like *while*, except that the code is run just before the the condition is evaluated. This structure is used when you want the code inside your block to run at least once before you check the condition.

Example:

```
do  {
  digitalWrite(13,HIGH);
  delay(100);
  digitalWrite(13,HIGH);
  delay(100);
  sensorValue = analogRead(1);
} while (sensorValue < 512);
```

break

This term lets you leave a loop and continue the execution of the code that appears after the loop. It's also used to separate the different sections of a *switch case* statement.

Example:
```
// blink LED while sensor is below 512
do  {
   // Leaves the loop if a button is pressed
   if (digitalRead(7) == HIGH)
     break;
   digitalWrite(13,HIGH);
   delay(100);
   digitalWrite(13,HIGH);
   delay(100);
   sensorValue = analogRead(1);
} while (sensorValue < 512);
```

continue

When used inside a loop, *continue* lets you skip the rest of the code inside it and force the condition to be tested again.

Example:
```
for (light = 0; light < 255; light++)
{
  // skip intensities between 140 and 200
  if ((x > 140) && (x < 200))
    continue;
  analogWrite(PWMpin, light);
  delay(10);
}
```

return

Stops running a function and returns from it. You can also use this to return a value from inside a function.

For example, if you have a function called *computeTemperature()* and you want to return the result to the part of your code that invoked the function you would write something like:
```
int computeTemperature() {
    int temperature = 0;
    temperature = (analogRead(0) + 45) / 100;
    return temperature;
}
```

ARITHMETIC AND FORMULAS

You can use Arduino to make complex calculations using a special syntax. + and − work like you've learned in school, and multiplication is represented with an * and division with a /.

There is an additional operator called "modulo" (%), which returns the remainder of an integer division. You can use as many levels of parentheses as necessary to group expressions. Contrary to what you might have learned in school, square brackets and curly brackets are reserved for other purposes (array indexes and blocks, respectively).

Examples:

```
a =  2 + 2;
light = ((12 * sensorValue) - 5 ) / 2;
remainder = 3 % 2; // returns 2 because 3 / 2 has remainder 1
```

COMPARISON OPERATORS

When you specify conditions or tests for *if*, *while*, and *for* statements, these are the operators you can use:

== equal to
!= not equal to
< less than
> greater than
<= less than or equal to
>= greater than or equal to

BOOLEAN OPERATORS

These are used when you want to combine multiple conditions. For example, if you want to check whether the value coming from a sensor is between 5 and 10, you would write:

```
if ((sensor => 5) && (sensor <=10))
```

There are three operators: and, represented with **&&**; or, represented with **||**; and finally not, represented with **!**.

COMPOUND OPERATORS

These are special operators used to make code more concise for some very common operations like incrementing a value.

For example, to increment *value* by 1 you would write:

```
value = value +1;
```

but using a compound operator, this becomes:

```
value++;
```

increment and decrement (−− and ++)

These increment or decrement a value by 1. Be careful, though. If you write *i++* this increments *i* by 1 and evaluates to the equivalent of *i+1*; *++i* evaluates to the value of *i* **then** increments *i*. The same applies to −−.

+= , −=, *= and /=

These make it shorter to write certain expressions. The following two expressions are equivalent:

```
a = a + 5;
a += 5;
```

INPUT AND OUTPUT FUNCTIONS

Arduino includes functions for handling input and output. You've already seen some of these in the example programs throughout the book.

pinMode(pin, mode)

Reconfigures a digital pin to behave either as an input or an output.

Example:
```
pinMode(7,INPUT); // turns pin 7 into an input
```

digitalWrite(pin, value)

Turns a digital pin either on or off. Pins must be explicitly made into an output using *pinMode* before *digitalWrite* will have any effect.

Example:
```
digitalWrite(8,HIGH); // turns on digital pin 8
```

int digitalRead(pin)

Reads the state of an input pin, returns HIGH if the pin senses some voltage or LOW if there is no voltage applied.

Example:
```
val = digitalRead(7); // reads pin 7 into val
```

int analogRead(pin)
Reads the voltage applied to an analog input pin and returns a number between 0 and 1023 that represents the voltages between 0 and 5 V.

Example:
```
val = analogRead(0); // reads analog input 0 into val
```

analogWrite(pin, value)
Changes the PWM rate on one of the pins marked PWM. *pin* may be 11,10, 9, 6, 5, 3. *value* may be a number between 0 and 255 that represents the scale between 0 and 5 V output voltage.

Example:
```
analogWrite(9,128); // Dim an LED on pin 9 to 50%
```

shiftOut(dataPin, clockPin, bitOrder, value)
Sends data to a *shift register*, devices that are used to expand the number of digital outputs. This protocol uses one pin for data and one for clock. *bitOrder* indicates the ordering of bytes (least significant or most significant) and *value* is the actual byte to be sent out.

Example:
```
shiftOut(dataPin, clockPin, LSBFIRST, 255);
```

unsigned long pulseIn(pin, value)
Measures the duration of a pulse coming in on one of the digital inputs. This is useful, for example, to read some infrared sensors or accelerometers that output their value as pulses of changing duration.

Example:
```
time = pulsein(7,HIGH); // measures the time the next
                        // pulse stays high
```

TIME FUNCTIONS
Arduino includes functions for measuring elapsed time and also for pausing the sketch.

unsigned long millis()
Returns the number of milliseconds that have passed since the sketch started.

Example:
```
duration = millis()-lastTime; // computes time elapsed since "lastTime"
```

delay(ms)
Pauses the program for the amount of milliseconds specified.

Example:
```
delay(500); // stops the program for half a second
```

delayMicroseconds(us)
Pauses the program for the given amount of microseconds.

Example:
```
delayMicroseconds(1000); // waits for 1 millisecond
```

MATH FUNCTIONS
Arduino includes many common mathematical and trigonometric functions:

min(x, y)
Returns the smaller of x and y.

Example:
```
val = min(10,20); // val is now 10
```

max(x, y)
Returns the larger of x and y.

Example:
```
val = max(10,20); // val is now 20
```

abs(x)
Returns the absolute value of x, which turns negative numbers into positive. If x is 5 it will return 5, but if x is −5, it will still return 5.

Example:
```
val = abs(-5); // val is now 5
```

constrain(x, a, b)
Returns the value of x, constrained between a and b. If x is less than a, it will just return a and if x is greater than b, it will just return b.

Example:
```
val = constrain(analogRead(0), 0, 255); // reject values bigger than 255
```

map(value, fromLow, fromHigh, toLow, toHigh)
Maps a value in the range *fromLow* and *maxLow* to the range *toLow* and *toHigh*. Very useful to process values from analogue sensors.

Example:
```
val = map(analogRead(0),0,1023,100, 200); // maps the value of
                                           // analog 0 to a value
                                           // between 100 and 200
```

double pow(base, exponent)
Returns the result of raising a number (*base*) to a value (*exponent*).

Example:
```
double x = pow(y, 32); // sets x to y raised to the 32nd power
```

double sqrt(x)
Returns the square root of a number.

Example:
```
double a = sqrt(1138); // approximately 33.73425674438
```

double sin(rad)
Returns the sine of an angle specified in radians.

Example:
```
double sine = sin(2); // approximately 0.90929737091
```

double cos(rad)
Returns the cosine of an angle specified in radians.

Example:
```
double cosine = cos(2); // approximately -0.41614685058
```

double tan(rad)
Returns the tangent of an angle specified in radians.

Example:
```
double tangent = tan(2); // approximately -2.18503975868
```

RANDOM NUMBER FUNCTIONS

If you need to generate random numbers, you can use Arduino's pseudo-random number generator.

randomSeed(seed)

Resets Arduino's pseudorandom number generator. Although the distribution of the numbers returned by *random()* is essentially random, the sequence is predictable. So, you should reset the generator to some random value. If you have an unconnected analog pin, it will pick up random noise from the surrounding environment (radio waves, cosmic rays, electromagnetic interference from cell phones and fluorescent lights, and so on).

Example:

```
randomSeed(analogRead(5)); // randomize using noise from pin 5
```

long random(max)
long random(min, max)

Returns a pseudorandom *long* integer value between *min* and *max − 1*. If min is not specified, the lower bound is 0.

Example:

```
long randnum = random(0, 100); // a number between 0 and 99
long randnum = random(11);     // a number between 0 and 10
```

SERIAL COMMUNICATION

As you saw in Chapter 5, you can communicate with devices over the USB port using a serial communication protocol. Here are the serial functions.

Serial.begin(speed)

Prepares Arduino to begin sending and receiving serial data. You'll generally use 9600 bits per second (bps) with the Arduino IDE serial monitor, but other speeds are available, usually no more than 115,200 bps.

Example:

```
Serial.begin(9600);
```

Serial.print(data)
Serial.print(data, encoding)

Sends some data to the serial port. The encoding is optional; if not supplied, the data is treated as much like plain text as possible.

Examples:

```
Serial.print(75);        // Prints "75"
Serial.print(75, DEC);   // The same as above.
Serial.print(75, HEX);   // "4B" (75 in hexadecimal)
Serial.print(75, OCT);   // "113" (75 in octal)
Serial.print(75, BIN);   // "1001011" (75 in binary)
Serial.print(75, BYTE);  // "K" (the raw byte happens to
                         // be 75 in the ASCII set)
```

Serial.println(data)
Serial.println(data, encoding)

Same as *Serial.print()*, except that it adds a carriage return and linefeed (\r\n) as if you had typed the data and then pressed Return or Enter.

Examples:

```
Serial.println(75);        // Prints "75\r\n"
Serial.println(75, DEC);   // The same as above.
Serial.println(75, HEX);   // "4B\r\n"
Serial.println(75, OCT);   // "113\r\n"
Serial.println(75, BIN);   // "1001011\r\n"
Serial.println(75, BYTE);  // "K\r\n"
```

int Serial.available()

Returns how many unread bytes are available on the Serial port for reading via the *read()* function. After you have *read()* everything available, *Serial.available()* returns 0 until new data arrives on the serial port.

Example:

```
int count = Serial.available();
```

int Serial.read()

Fetches one byte of incoming serial data.

Example:

```
int data = Serial.read();
```

Serial.flush()

Because data may arrive through the serial port faster than your program can process it, Arduino keeps all the incoming data in a buffer. If you need to clear the buffer and let it fill up with fresh data, use the *flush()* function.

Example:

```
Serial.flush();
```

Appendix D/Reading Schematic Diagrams

So far, we have used very detailed illustrations to describe how to assemble our circuits, but as you can imagine, it's not exactly a quick task to draw one of those for any experiment you want to document.

Similar issues arise, sooner or later, in every discipline. In music, after you write a nice song, you need to write it down using musical notation.

Engineers, being practical people, have developed a quick way to capture the essence of a circuit in order to be able to document it and later rebuild it or pass it to somebody else.

In electronics, *schematic diagrams* allow you to describe your circuit in a way that is understood by the rest of the community. Individual components are represented by symbols that are a sort of abstraction of either the shape of the component or the essence of them. For example, the capacitor is made of two metal plates separated by either air or plastic; therefore, its symbol is:

Another clear example is the inductor, which is built by winding copper wire around a cylindrical shape; consequently the symbol is:

The connections between components are usually made using either wires or tracks on the printed circuit board and are represented on the diagram as simple lines. When two wires are connected, the connection is represented by a big dot placed where the two lines cross:

This is all you need to understand basic schematics. Here is a more comprehensive list of symbols and their meanings:

RESISTOR   CAPACITOR   THERMISTOR   LDR LIGHT SENSOR

DIODE   LED   PUSHBUTTON   POTENTIOMETER

You may encounter variations in these symbols (for example, both variants of resistor symbols are shown here). See en.wikipedia.org/wiki/ Electronic_symbol for a larger list of electronics symbols. By convention, diagrams are drawn from left to right. For example, a radio would be drawn starting with the antenna on the left, following the path of the radio signal as it makes its way to the speaker (which is drawn on the right).

The following schematic describes the push-button circuit shown earlier in this book:

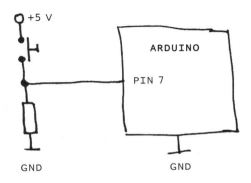

Index